Personal
Computers
for Scientists

Personal Computers for Scientists

A Byte at a Time

Glenn I. Ouchi
BREGO Research

American Chemical Society
WASHINGTON, DC   1987

Library of Congress Cataloging-in-Publication Data
Ouchi, Glenn I., 1949–
    Personal computers for scientists.

    Includes index.

    1. Microcomputers. 2. Science—Data processing.

    I. Title.

QA76.5.0924   1986   004.16   86–25846
ISBN 0–8412–1000–4
ISBN 0–8412–1001–2 (pbk.)

About the Author

Glenn I. Ouchi is founder and president of BREGO Research, founder of the Laboratory PC Users Group, and editor of *Laboratory PC User*, a monthly newsletter for personal computer (PC) users in chemical laboratories. He received his B.S. in chemistry from the University of California, Los Angeles; his M.S. in organic chemistry from the University of Minnesota while working with Edward Leete; and his Ph.D. in computer applications in organic

chemistry from the University of California, Santa Cruz, while working with W. Todd Wipke. His thesis project was the development of the first version of XENO, a program that predicts the plausible metabolites of xenobiotic compounds in mammals. He has published papers on subjects ranging from the metabolism of DDT analogs by microorganisms to chemical applications of artificial intelligence.

Dr. Ouchi's professional career includes work in the pharmaceutical analysis group at SRI International, developing software for gas chromatography–mass spectrometry (GC–MS) applications at Finnegan MAT, and developing software for chromatographic applications at Spectra-Physics and Nelson Analytical. He has designed many commercial software packages used in GC–MS, gel permeation chromatography, simulated distillation, and chromatographic data handling. He wrote the first IBM PC-based chromatographic data system and was product manager for the data system while at Nelson Analytical.

Dr. Ouchi has been active in promoting the use of PCs in the chemical laboratory. He has taught the American Chemical Society

(ACS) short courses, "Personal Computers for Chemists" and "Interfacing Laboratory Instruments to PCs", and has given many seminars at ACS national meetings on PC use. His articles on PC applications for chemists have appeared in *PC World, PC Week, Analytical Chemistry, American Laboratory,* and *R&D.*

To Gail, Rochelle, Bennett, Jean, my mother, and
the memory of my father

Contents

Preface

This book is designed to help you select and use a personal computer (PC) to solve problems in your work. This book will also be helpful if you already have a PC that performs certain functions and you are looking for other applications. This book assumes you have had little or no experience with computers. It will dispel the myth that computers and the people who program computers have a magical quality.

Computers are just dumb machines (really no smarter than can openers), and the people who write programs to make computers operate are ordinary people. If there is magic in a computer, it is because computers have the ability to store, retrieve, and perform mathematical calculations on thousands of lines of text or numbers in a few seconds with no chance of error. Computers can store text and numbers and apply them in another form or at another time at very high speed. Computers can take the time-consuming mechanical aspects out of mundane tasks like drawing graphs or writing articles and leave time for you to perfect the content. These capabilities make the computer a perfect tool for your work and make you more productive.

This book is software-oriented because application software allows you to solve your problems, organize your information, and plot your data. All too often I see scientists choosing computers only for their hardware performance (fast central processing unit [CPU] or large available memory) without any thought given to what software is available. Then the scientists are forced to spend years developing or waiting for the development of the necessary software to perform an application that could have been solved using "off-the-shelf" software available for a different computer. This common error in computer selection is a terrible waste of time and resources. This book stresses the strategy of finding application software that will solve your problems and will enable you to select the computer hardware that best runs the software.

This book is divided into three sections. The first section provides a basic introduction to computer software and hardware.

The second section introduces six types of application software that can be used to analyze, display, communicate, and organize scientific data. All of the software and hardware described is off-the-shelf, so you can purchase it and immediately solve your problems. Each chapter includes examples of program use in scientific applications. Each application shows innovative ways to use a PC. These applications present concrete examples of the powerful capabilities of a PC and show how quickly a PC can be productive for you.

The third section of the book discusses data communication and interconnecting your PC with the real world.

Some chapters include example programs and program "templates" for popular programs that run the IBM PC, PC/XT, PC/AT, and most IBM PC clones. The program and template listings are provided in the appendix.

I would like to thank the many people who made this book possible, in particular Keith Belton, Paula Bérard, and Robert Johnson of the ACS Books Department.

GLENN I. OUCHI
BREGO Research
San Jose, CA 95124

June 1986

Section ONE

Personal Computers for the Chemical Laboratory

Science is organized knowledge
—Herbert Spencer in *Education*

This section concentrates on the basic parts of a personal computer (PC), your information organization machine for laboratory work. The PC provides the power to compute, organize, and communicate. With a PC, mountains of data can be transformed into information and knowledge.

Chapter

Introduction to Personal Computers

elcome to the Information Age. We are rapidly shifting from an industry-based to an information-based society. As a scientist, you are already a veteran of the Information Age. You spend a major part of your day working with information rather than actual objects or things. You perform tests to observe a physical phenomenon (Figure 1.1), and through experimentation you capture raw data by using sensors or detectors that measure the phenomenon. This raw data is then analyzed and turned into information from which you can make decisions about the phenomenon. You are trained to design, execute, observe, and analyze experiments that are information-gathering expeditions. Other than the execution of the experiment, where you probably will work with objects or things, the balance of your time is spent working with information to design, observe, analyze, and communicate your results to others so you or they can take action.

Data acquisition   Analysis   Action

Phenomenon ⟶ Raw data ⟶ Information ⟶ Result

Figure 1.1   Data acquisition, analysis, and action.

A personal computer (PC) can provide the tools to perform many of your information-handling needs more effectively. By keeping a data base of previous experiments, a PC can aid in new experiment design. A PC can be used to model an experiment and provide information on what should be expected as a result. A PC can be used in the observation and analysis stages of an experiment by acquiring raw data and aiding the detailed analysis that turns raw data into timely information. The information or knowledge gleaned from the

1000–4/87/0003$06.00/1 © 1987 American Chemical Society

raw data can also be easily maintained, manipulated, correlated, and communicated through PC use. The PC has the potential to be the major tool used to design and perform experiments, capture results, analyze data, organize information, and communicate information to others.

With the proper interfaces, a PC can also be used to control a laboratory instrument, robot, or experiment and collect the raw data. In this application, the computer can perform exact repetitive actions that ensure the experiment is performed with the same techniques each time. Thus, the PC frees us from mindless tasks.

Revolutionary Development of the PC

Fewer than ten years ago, when I started using computers for chemical research, most computers were large, delicate, and extremely expensive machines. I worked on a mainframe computer that required an air-conditioned room, raised floors, and an army of maintenance and service personnel. At that time, even minicomputers took a large amount of space and required special training to operate and maintain. Because of the high cost of these computer systems, only small groups of people working in large institutions had access to a computer's power to organize information. I was able to use a national computing shared resource called the Stanford University Medical Experiment–Artificial Intelligence in Medicine (SUMEX–AIM). The computer had 512 kilobytes of random access memory (RAM), and I was allowed about 10 megabytes of disk storage. I worked on the computer through either a text or graphics terminal. Now, because computers have developed so rapidly, I have about the same computing power of that shared resource sitting on my desk for my own use.

In the past five years, the greatest revolution in information access has occurred, along with the capability to organize information with the development and distribution of the low-cost PC system. The computer's power to organize is now available to almost everyone. Because of the low cost of this computing power, new applications appear every day. Scientists can immediately take advantage of these low-cost computers.

Today, a PC with as much computing power as yesterday's mainframe and minicomputers can sit on a desk or laboratory bench. The portable models can go anywhere and can be operated by batteries in remote environments. Disk drives, printers, and other peripheral computer equipment have followed a similar revolutionary development path, and complete computer systems are now relatively inexpensive. Along with being low-cost, this equipment is also more reliable and simpler to use than its computing ancestors. These low-cost, reliable products have changed how computers are used and the economics of their application.

The revolutionary changes in computing power have occurred so fast in the past few years that applications that once would have only been attempted on large computers can now be performed routinely on a PC. The challenge then is to learn to use this new evolving tool to help solve daily problems.

What Is a PC?

Shown in Figure 1.2 is the IBM personal computer Advanced Technology Model, or PC/AT. It is small enough to fit on a desk or laboratory bench, requires no special power sources, and is relatively

Figure 1.2   The IBM PC/AT system including the main system unit housing the computer and disk drives, a graphics monitor, and a keyboard. (Photo courtesy of IBM.)

quiet in operation. In normal operation, the PC/AT does not require any special servicing or maintenance.

Hardware and Software

PCs have two main parts: hardware and software. The *hardware* is the constant part of the computer you can actually see, touch, and feel. If an analogy is made between a computer and a stereo system, the hardware is the speakers, amplifier, turntable, and wires connecting these parts together. *Software* is the variable part of the computer, the programs that make the computer display data, draw graphs, or entertain you with games. In the stereo analogy, software is like the records, tapes, or audio (compact) disks that can be played. With the same stereo (hardware), many kinds of records and music (software) like jazz, rock, or classical, can be played. Even when new records or tapes are produced, current hardware can play them.

Software: The Key to Productivity

Because extremely low-cost, highly reliable computer hardware has been widely available, the focus for problem-solving applications has switched to software. In the past, software was normally developed by the owner of the computer. The computer manufacturer supplied some basic software, called systems software, in the form of operating systems, programming languages, and some utilities, but the user had to write the software to perform applications. Using the computer–stereo analogy, this would mean you would have to record your own music. Previously, a small number of third-party software houses wrote and sold software. The cost of this software was extremely high (normally $50,000 and more) because of the high cost of development, little or no competition, and the extremely small potential market for a particular software product.

The product that has had the biggest effect on PC software development has been the IBM PC disk operating system (PC–DOS), which is an adaptation of the Microsoft disk operating system (MS–DOS). The IBM PC has drastically changed the economics of software development by setting a de facto standard for PC hardware and disk operating systems. Prior to the introduction of the IBM PC, few hardware standards existed. For competitive reasons, each computer manufacturer made computers that could not "play" the software written by their competitors. This situation was similar to stereo manufacturers making different stereos that could not play the same records, tapes, or compact disks. Each time a new stereo model was introduced, new records would have to be recorded. Under these conditions, the records or software were very expensive because each was essentially custom-made in low volumes. Also, because the potential market for each software package was very small, very few companies would take the risk to write software for a specific computer.

Having a hardware and operating system standard has changed software development dramatically. The potential market for a software product that runs on the IBM PC or one of its clones is immense. Now, more than 5 million IBM PCs are in operation. Add to that number the equally large number of IBM PC look-alike computers called PC clones, and more than 10 million potential buyers of software result if the application runs under PC–DOS or MS–DOS. The immense popularity of the IBM PC is also due to "open architecture". *Open architecture* means IBM invites anyone to create add-on software and hardware products by publishing the specifications for IBM's add-on hardware slots. The MS–DOS operating system is available to other vendors, so software written for the IBM PC can also run on computers manufactured by other companies that use MS–DOS.

The most publicized software success that took advantage of the de facto hardware standard is the integrated software program, Lotus 1–2–3. Lotus Development Corporation's first products, Lotus 1–2–3 and Symphony, have now sold much more than 1.5 million copies with list prices of $495 and $695, respectively. This extremely successful software product and many more like it would never have been developed, nor sold for such an attractive price, if a huge potential market for the product did not exist. These products took full advantage of the IBM PC's fully integrated problem-solving system, including fast on-screen graphics and full support for many printers and plotters.

The Upward Spiral of Product Development Features

The IBM PC's market potential attracts companies to produce new products for the computer. If the new product must compete with existing products, the new product must either provide all the features of the previous products at a lower price or provide more features at the same or lower price. The result to the consumer is more features per dollar. As more and better software is developed for more applications, more computers are sold, and the market potential increases. The computer industry has produced a fantastic upward spiral of better products for lower cost. Weak products in this highly competitive market will not exist long, and thus only the best products (the most features for the lowest price per feature) will survive. This upward spiral also applies to peripheral computer accessories such as disk drives, printers, plotters, and monitors.

We are living in an extremely interesting time in computing history. Never before has computing power been available to so many people. And the opportunity to use this computing power to solve your problems has never been greater because access to so many excellent software tools is available. The great number of excellent software tools differentiates the PC computing environment from the environment of minicomputers and mainframe computers.

Solving Problems Using a Computer

Brainware

This computing power does not just pop out of a new computer's shipping box nor the fancy shrink-wrapped manuals and disks of newly purchased software. You will have to work to include a computer in a problem-solving process. Three major components are required to solve problems using computers. Hardware and software have already been mentioned. The third, knowledge and training on hardware and software use, is just as important. One training company, KnowHow, Inc., of San Francisco, has coined the term *brainware* for this third component.

Brainware is gained just like any other skill, such as learning to ride a bicycle, play tennis, or design the synthesis of life-saving compounds. Hard work, discipline, and energy are needed to master the use of a computer for problem solving. No matter how good the computer hardware or how "user friendly" the software, it takes time to master a new talent. And, as in learning to ride a bicycle, you must experiment as you learn in order to make the fastest progress. To learn about computing on a PC is very advantageous because while you experiment with the PC, you cannot interfere with the work of others, as you would if you were using a shared minicomputer or mainframe computer.

Gaining Computer Knowledge

As in all new endeavors, initial progress will seem slow, but with constant effort you will reach your goal. Reading this book is one way to gain brainware, and many other ways to gain computer knowledge exist.

Read. Read everything you can (Figure 1.3). Reading is the most cost-effective way to obtain information you will need to feel comfortable with computers. For current information, weekly, bimonthly, and monthly magazines cover the PC industry. Soon, a daily publication will provide up-to-the-minute information. These publications are available both free and via subscription. Here is my list of favorites:

PC Week is a Ziff Davis publication. This weekly is free to qualified readers or available via subscription. PC Week is packed with timely information about products, excellent product reviews, and the best weekly columns in the computer industry.

PC Magazine is another Ziff Davis publication. This bimonthly magazine requires a paid subscription and has excellent product

Figure 1.3 Example of the many magazines, books, and journals published that help users.

reviews. These reviews have more depth than those in *PC Week* and sometimes span hundreds of pages. Articles are also excellent and hit specific application areas in each issue such as medicine and law. *PC Magazine* has by far the best product evaluations; complete issues have been dedicated to reviews of more than 200 printers and more than 50 data base management systems over a period of three months in the "Operation Database" series.

PC World, BYTE, Software News, and *Mini–Micro Systems* are all monthly publications. *PC World* is a good magazine for the beginning computer user; it is published by PC World and requires a paid subscription. PC World also publishes *Mac World,* which supports the Apple Macintosh computer. Both magazines provide a special section called "Getting Started", just for first-time users.

BYTE is the granddaddy of PC magazines, and it provides insights to all aspects of personal computing. *BYTE* is published by McGraw–Hill and requires a paid subscription.

Software News covers software for PCs, minicomputers, and mainframe computers. It is published by Sentry and is free to qualified individuals.

Mini–Micro Systems is a more general publication covering PCs and minicomputers. *Mini–Micro Systems* is published by McGraw–Hill and is free to qualified subscribers.

Books on computers and computer applications are also readily available. Visit your company or community library or local bookstore to see the latest titles. Most magazines also review books.

Attend a Computer Course. Learning to use a computer requires concentrated effort. In most cases, because our daily jobs require our full attention, we rarely have time to learn about computers. A short course on computer usage can be very valuable because it will not only introduce you to computing but will also get you out of your daily routine for enough time to clearly evaluate the potential of using a computer in your problem solving. Introductory computer courses are offered by the American Chemical Society, other professional societies, and local colleges. Courses are also offered by many computer stores and professional training companies that specialize in computer training.

Join a Users Group. One of the fastest ways to learn about computers is to get involved in a local users group. Normally, a users group meets once or twice per month to discuss various aspects of computing. Some of the larger groups have invited speakers from computer software and hardware companies. Most users groups have subgroups with specific interests such as scientific computing or graphics. By participating in a users group, you will meet people with shared computing interests. Most groups are formed by users of the same computer, and the group will have a library of public domain software that is available to you. The best aspect of a users group is that first-hand information about computers and computer products can be obtained from people who have tried the products. This information is provided normally in a monthly newsletter (Figure 1.4). Even if you cannot attend meetings, the newsletter and access to the users group software library is well worth the low users group fee.

I have formed a users group focusing on chemical laboratory applications of PCs. We have a monthly newsletter, a bulletin board system, and a large library of public domain software. If you are interested in joining, write: LAB–PC Users Group, c/o Glenn Ouchi, 5989 Vista Loop, San Jose, CA 95124, or call (408) 723-0947.

Subscribe to a National Computer Network or Go On-Line to Some Local Bulletin Boards. You can use a number of national computer networks via telephone by purchasing a modem and some communications software and paying a subscription fee. These information services offer everything from *Chemical Abstracts* to stock quotations. These services also have *electronic mail,* so you can send and receive messages from other users on the network. Particularly helpful are special-interest group areas where a message or question about almost any subject can be left and someone will provide ideas on a solution. Bulletin board systems can be accessed via a telephone line and a modem. Bulletin boards provide communications ability similar to the national networks but on a much smaller scale; therefore, a unique method of communication between users of PCs is provided. Most bulletin boards have electronic mail and freeware or shareware programs (*see* Chapter 7). Chapter 10 gives more information about on-line information networks and bulletin board systems.

Talk to Friends about Computers. Your friends, even if they are in different professions, can provide valuable knowledge about computers if they are using them. Two of my friends, a lawyer and an investment dealer, use computers in their work and have given me many ideas about how I could use my computer system more

Laboratory PC User

Newsletter for the Laboratory PC Users Group

Volume 1, Number 3	August 1986	$36.00 per Year

Managing Your Instrument Data

Using a PC in the laboratory to solve data management problems requires planning and wise implementation. But don't spend a long time selecting your first data management system. Find an easy one to use and get your feet wet as soon as you can!

The Big Five
Software Applications

Graphics / Communications

Spreadsheets / Data Management / Word Processing

Figure 1. The Big Five PC Software Applications.

Inside This Issue:

Data Management is one of the "Big Five" Horizontal Software applications, Figure 1. The others are word processing, spreadsheets, communications, and graphics. A combination of these horizontal software applications, your laboratory data, some interfacing software and lots of Brainware will create an integrated solution to most of your laboratory computing problems, Figure 2. You should think of these horizontal software products as building blocks for your solution. In each of these five major types of software there are literally hundreds of different programs you can select from. Your job is to select the package which fits your application the best.

Lab Data Solution Central

At the center of your computer solution is your laboratory data. This data can come from instruments or you can acquire it directly from sensors positioned on your experiment. Data is central to your PC solution. If the raw data is acquired poorly, at too fast or slow a rate or inaccurately converted into a digital form, no matter how perfectly the data is processed, the application will provide erroneous results. Your first task is to insure your raw data is accurately acquired. After data is acquired, it is usually placed in a file on a disk for more permanent storage. Once your data is on disk, other application programs can read the data for further processing, graphing and reporting. Between the laboratory data and the application programs may be one or more "interfacing" programs or procedures like those

Continued on Page 3

Figure 1.4 Users groups publish helpful information in monthly newsletters. They also offer public domain software and information via electronic bulletin boards.

efficiently. When you first start using your computer, calling or talking with someone, even with seemingly small questions, is very helpful. These seemingly trivial questions can easily be answered by a person with a little computing experience, yet to find the answer in a manual, reference guide, book, or magazine article would take an unknown amount of time. Even if the person cannot answer your question, he or she may know where to find the answer.

To gain knowledge about computers will require some effort, but once the first hurdle is cleared, computers become extremely valuable in solving problems and organizing information.

Application Software

Basically, a computer has the capability to accept information in the form of numbers or words (input), efficiently manipulate the information (process, rearrange, display, and compute), and then report the results (output). PC software provides the tools to perform these tasks. Without software, a computer is just a package of metal, plastic, and silicon.

Innovative Software Is the Key

Paralleling the exciting improvements in computer hardware have been vast improvements in computer software, such as full-screen editing, color graphics, direct control with pointers (mice), integrated applications, and interactive graphics. The new wave of low-cost application software for PC systems makes computer operation easy for anyone. Special programming languages are no longer needed to perform productive work. The improved software has allowed personal computing to expand rapidly and is available for anyone who needs to organize information for reporting, planning, or decision making. Scientists especially can take advantage of the various functions to organize, analyze, and report information for their research.

Major Types of Application Software

In the rest of this chapter and the chapters that follow, the major types of PC software available today will be explored. A quick introduction will be presented here, then a full chapter will be devoted to each type. Most of the thousands of application programs that have been developed for PCs fall into five broad categories: data communication, spreadsheets, graphics, data base management, and word processing. Nearly all applications known for computers can be broken down into modules that use the features of these main program categories. You should be familiar with the capabilities of each type of program, so when faced with a problem, you can quickly design a solution.

Data Communications. Data communications software allows communication with a broad range of other computers. This type of software is particularly important to scientists because raw data can be acquired either directly from experiments or from another computer that is built into a laboratory instrument. This type of software also allows communication with larger computers that contain data bases and scientific literature. Many scientists can justify the expense of a PC because of the time saved in performing literature searches alone. Finally, this software is used when sending results to another computer that is storing the data or will correlate the results with the results of others.

Spreadsheets. Spreadsheet programs allow you to use a grid of columns and rows similar to pages of graph paper in a notebook. Each intersection of a row and column can hold information and is called a *cell.* Each cell is filled with information such as experimental results, instrument readings, and calibration information. Cells can contain text for titles, column, or row headings; numbers for results; or equations for computing new values from data. Once set up, the data reduction or transformation equations on a spreadsheet form a template for a report that, by simply changing the original data, will generate a new report (Figure 1.5). You can also ask "what-if" questions by changing variables and thus instantly see the result. To get output from a spreadsheet, simply print the contents of the display.

Graphics. Analysis of scientific information is particularly suited for graphic analysis. Graphics software allows you to see your data displayed in many formats. The graph in Figure 1.6 is easier to interpret than the columns of numbers. The speed of generating graphic displays is also important because plotted data can be seen instantly, then redisplayed after a few variables have been changed. This type of interactive graphics capability is extremely valuable for data analysis.

	A	B	C	D	E	F
1	Methane Samples					
2						
3	Submitter	Amount	Area	Res. Factor		Submitter
13	S. Kofax	450	4500	10		
14	N. Larker	650	6500	10		
15	G. Hodges	480	4800	10		
16	J. Robinson	610	6100	10		
17	P. Reese	230	2300	10		
18	J. Gilliam	340	3400	10		
19	C. Neal	900	9000	10		
20	D. Demeter	350	3500	10		
21	F. Howard	410	4100	10		
22						
23						
24	Data Base Stats					
25	Mean	468.8235	4688.235			
26	Standard Dev	160.9121	1609.121			
27	% Std Dev	34.32255	34.32255			
28	High	900	9000			
29	Low	230	2300			

Figure 1.5   Spreadsheet screen containing template report.

Data Base Management.   Data base managers allow you to store data and text and then retrieve the data in many types of report formats. These programs allow you to create data bases, enter the data through data entry screens (Figure 1.7), and then either through report generators or query commands, review or report the data. Data base managers give flexibility to review, compare, analyze, and report data in many different formats without re-entering data for a particular application. New application programs using the same data from a data base are easier to design and implement than starting from scratch.

Word Processing.   Word processing software transforms a PC into a super typewriter. You compose your text (such as reports, experimental results, grant proposals, or letters) on the video monitor instead of typing on paper, so corrections are easy to make. You can delete, insert, and rearrange words, paragraphs, and blocks of text. The text is displayed on the screen in the same form it will be typed on the paper. Spelling checker modules help to ensure that all words in any document are correctly spelled. The document can be printed in a number of formats.

A	B
182	8976
183	8972
184	9525
185	11432
186	13497
187	15748
188	19570
189	26026
190	44834
191	82390
192	122702
193	166432
194	205905
195	236990
196	250473
197	229006
198	196136
199	166873
200	138234
201	112675

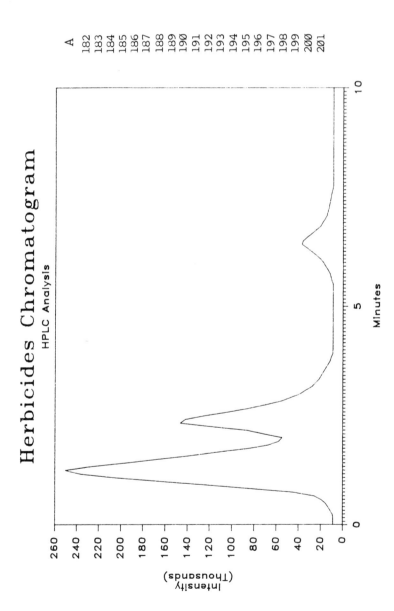

Figure 1.6   Graphic display of data compared with printed data.

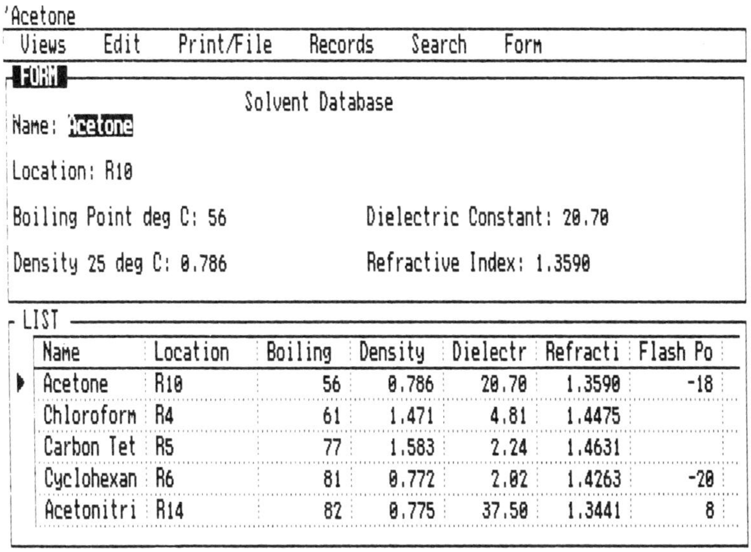

```
'Acetone
 Views    Edit    Print/File    Records    Search    Form
▐FORM▌─────────────────────────────────────────────────
                     Solvent Database
 Name: ▐Acetone▌

 Location: R10

 Boiling Point deg C: 56           Dielectric Constant: 20.70

 Density 25 deg C: 0.786           Refractive Index: 1.3590

┌ LIST ──────────────────────────────────────────────────
│   Name        Location   Boiling   Density   Dielectr   Refracti   Flash Po
│ ▶ Acetone      R10            56     0.786     20.70      1.3590       -18
│   Chloroform   R4             61     1.471      4.81      1.4475
│   Carbon Tet   R5             77     1.583      2.24      1.4631
│   Cyclohexan   R6             81     0.772      2.02      1.4263       -20
│   Acetonitri   R14            82     0.775     37.50      1.3441         8
```

Figure 1.7   Data base entry screen.

Integrated Application Programs

Integrated programs combine two or more of the major program types. The resulting program has added utility because data can be manipulated with many different types of software tools without having to enter the data again. For example, data can be entered into a spreadsheet program and then plotted (Figure 1.8) or sorted (Figure 1.9). Integrated application programs give more flexibility for developing a solution to problems. The more popular integrated programs, like Lotus' Symphony and Ashton–Tate's Framework, integrate all of the basic functions in one package. This approach can be very advantageous for a new PC user because only one program has to be learned rather than five or six separate programs.

Hardware and Software Combination Application Systems

These complete turnkey application systems cover a wide range of applications from computer-aided design (CAD) of printed circuit boards and circuit schematics to acquiring and processing chemical chromatography data. These systems are a combination of hardware (such as graphics plotters and instrument interfaces) and software that

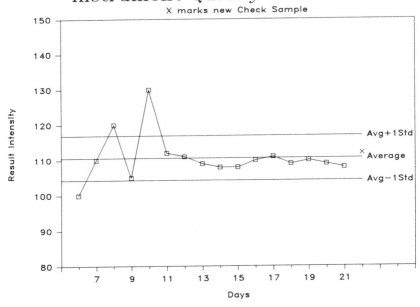

Figure 1.8 Plotted data from Lotus 1–2–3.

	K	L	M	N	O	P
6						
7	Ref no.	Rec Date	Submitted by	Analysis	Analyte	Raw Value
8	2003	8307.05	Calibration	GCHROM	ethane	120132
9	2006	8307.12	Calibration	GCHROM	ethane	124098
10	2009	8307.19	Calibration	GCHROM	ethane	118043
11	2012	8307.26	Calibration	GCHROM	ethane	112043
12	2015	8308.02	Calibration	GCHROM	ethane	107021
13	2018	8308.09	Calibration	GCHROM	ethane	100103
14	2002	8307.05	Calibration	GCHROM	methane	150032
15	2005	8307.12	Calibration	GCHROM	methane	160054
16	2008	8307.19	Calibration	GCHROM	methane	145045
17	2011	8307.26	Calibration	GCHROM	methane	155076
18	2014	8308.02	Calibration	GCHROM	methane	130102
19	2017	8308.09	Calibration	GCHROM	methane	120034
20	2001	8307.05	Calibration	GCHROM	propane	230145
21	2004	8307.12	Calibration	GCHROM	propane	234024
22	2007	8307.19	Calibration	GCHROM	propane	228023
23	2010	8307.26	Calibration	GCHROM	propane	232076
24	2013	8308.02	Calibration	GCHROM	propane	227032
25	2016	8308.09	Calibration	GCHROM	propane	225021

Figure 1.9 Sorted data from Lotus 1–2–3.

perform a specified task once performed only by more expensive dedicated systems. Compared with custom-designed equipment, these systems are frequently only one-tenth or even one one-hundredth the price. Of course, when the specific application is not being performed, the IBM PC can run other software and applications.

Chapter

Personal Computer Hardware

efore continuing the discussion of application software, the major computer hardware components and terms will be introduced. As in all fields, a large number of new terms and acronyms must be mastered. This chapter will give you enough information about computer hardware to understand what parts of the computer are used when application software is executed.

The physical equipment that makes up a PC is called *hardware*. This chapter will describe the different types of hardware that comprise a computer system. All computers have these same basic components.

Printed Circuit Boards

Whenever you look inside a computer, no matter how large (mainframe) or small (PC), you will find one or more objects like those in Figure 2.1. The dark rectangular packages are *integrated circuits,* which are also called ICs or chips. Integrated circuits are so named because they combine in a single package the features that once required a number of transistors, capacitors, resistors, and other electronic components. Today, thousands of these chips perform various computing jobs.

A chip by itself is not very valuable. It must be supplied with power and also send and receive data or instructions from other chips in the computer. This communication and power distribution is provided by the thin metallic tracks printed into the fiberglass circuit board. Most printed circuit boards have tracks on both sides of the board; more complex boards have three or four layers of tracks that interconnect the chips. The multilayered boards have interconnecting tracks stacked on each other like the parts of a sandwich. The major task in

1000–4/87/0021$07.00/1 © 1987 American Chemical Society

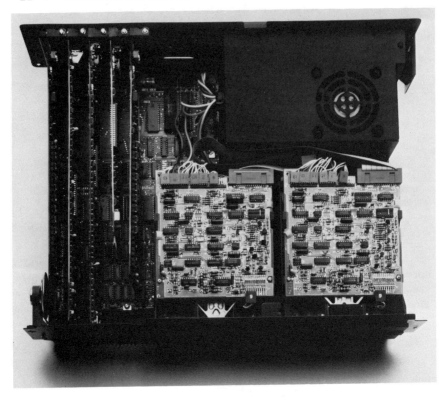

Figure 2.1   The main system board of the IBM PC. (Photo courtesy of IBM.)

designing a computer is to select the proper integrated circuits to perform the task and then lay out the circuit board on which the components can be mounted.

Central Processing Unit

The *central processing unit* (CPU) is the portion of the computer that performs calculations. The CPU performs all additions and comparisons and orchestrates where data is to be read from, stored, or sent. At the simplest level, the processor is a device with three memory locations. These memory locations are sometimes called *registers*. One location contains a command or instruction, and the other two locations contain data. According to the command, the processor adds, subtracts, or compares the contents of the data locations. With just these seemingly simple capabilities, the processor can perform almost any type of mathematics. If addition and subtraction can be done, then multiplication and division can be performed by a series of additions or subtractions. Because addition, subtraction, multiplication, and division are possible, then squares, square roots, and logarithms can be performed. Because text characters are represented in the computer as numbers, these characters can be compared, combined, or deleted by instructions from the processor.

The processor acts on the instructions given by the software program. These instructions are provided to the CPU in a form called *machine language*. There are many brands of CPUs. Some of the more popular brands are shown in Table 2.1. Each CPU has a specific list of functions or instructions. A program is a list of these instructions that tell the CPU exactly what calculations to perform. The number of instructions the CPU can perform and the speed at which it can perform them define the speed and capability of the CPU. The instructions are specific for each CPU; therefore, programs written for one CPU cannot be performed on a different CPU. To extend a CPU's useful life, software-compatible families of CPUs have been developed. An example is the Intel 8088, 8086, and 80286 family of CPUs. These CPUs have a common set of instructions that all can perform. To improve performance, more instructions or the ability to perform the functions faster can be added.

Memory

Memory in a computer is the hardware where program instructions (e.g., the program instructions to perform the program Lotus 1–2–3) and data (e.g., the data entered into a program) are stored while the CPU performs the calculations. Two types of memory are used in a computer. *Random access memory* (RAM), better described as read–

Table 2.1   Popular CPUs for PCs

Name	Original Manufacturer	Computers That Use the CPU
Z–80	Zilog	Osborne, Kaypro
6800, 6809	Motorola	SS–50 Bus, Tandy
6502	Mostek	Apple II, IIe, IIc, Commodore 64, ATARI
6800	Motorola	Apple Macintosh, Hewlett Packard 9000, AT&T 7500, Commodore Amiga, ATARI ST
8088	Intel	IBM PC, Compacq, many compatibles
8086	Intel	AT&T 6300
80286	Intel	IBM PC/AT, many compatibles
80386	Intel	next generation of PCs

Measurement of Memory: Bytes and Bits

Memory is simply a large set of electronic switches that can either be on (electricity flowing through the switch) or off (no electricity flowing through the switch). These two states of the switch can be represented as *1* (on) and *0* (off). A single switch, called a binary digit, or bit, is not of much use alone because it can represent only two things. To represent a variety of things such as numbers, letters, and graphic shapes, a number of bits must be combined together.

Number of Bits	Number of Unique Representations
5	32
6	64
7	128
8	256

How many bits to cluster together is determined by the number of things the computer needs to represent. For example, the following items require 130–140 characters:

Characters To Be Represented	Number of Characters
Capital letters	26
Lower-case letters	26
Numbers	10
Punctuation and math symbols	23
Graphics characters	>20
Error checking	>10
Device control characters	>20
Total	135

Thus, 8 bits are needed in this example. A special code, the American Standard Code for Information Interchange, or ASCII (pronounced as'-key), was created to assign each letter, number, device control symbol, and punctuation mark a particular binary code.

Computer memory is thus divided into 8-bit clusters called *bytes*. The amount of memory in a computer is described in bytes. Normally, the number of bytes is measured in kilobytes, or K. One kilobyte of memory is actually 1024 bytes (2 raised to the 10th power). Because a byte through the ASCII code also represents a specific letter or number, a byte is also called a character. The ASCII codes for the letters *H* and *O* and the number *2* are shown here.

H 1001000   O 1001111   2 0110010

write memory, is used for the temporary storage of the executing program instructions and data. The CPU can read the program instructions to execute the program and write new data into RAM or replace data that is already there. This memory is termed volatile because if the electrical power is lost, the contents of the RAM will be lost.

Read-only memory (ROM) is also used to store program instructions. Because ROM can only be read, it is normally not used to store data. ROM is normally used to store unchanging program instructions like those needed when the computer is first turned on. In the IBM PC, ROM is used to store the "driver" instructions for many pieces of peripheral equipment like the disk drives and monitor cards. Program instructions in ROM are not volatile and are safe even when the power is off.

The amount of memory in a computer is measured in *kilobytes* (K); each kilobyte equals 1024 bytes. Because memory chips were expensive, early PCs had very little memory. The original Apple computer in 1977 had only 4K memory, barely enough to execute a simple game. Today, because memory chip prices have plummeted, 512K or 640K RAM PC systems are common. Figure 2.2 compares bytes with text pages.

Addressing Memory

When a program is executed, the contents of the computer's memory must be organized in some way to know which part of memory to use. Computer memory can be thought of as a wall filled with small post office boxes. To distinguish memory locations, each location has an *address.* The CPU knows which location in memory to read when a piece of data or the next program instruction is required. The amount of memory a computer can use is determined by how large an address it can read and write. Each CPU has a specific number of address lines that sets the maximum amount of addressable memory. For example, an 8-bit CPU with 16 address lines in its address bus can address 65,536 bytes, or 64K, different memory locations. The Intel 8088 processor in the IBM PC has 20 address lines; it can address 1,048,576 bytes, or 1024K or 1 megabyte of storage. The additional memory space can be used to load more data or programs.

If necessary, additional memory can be addressed by using a technique called memory *paging* or *bank switching.* This technique allows a CPU that can, for example, address only 64K to actually use 128K or more memory.

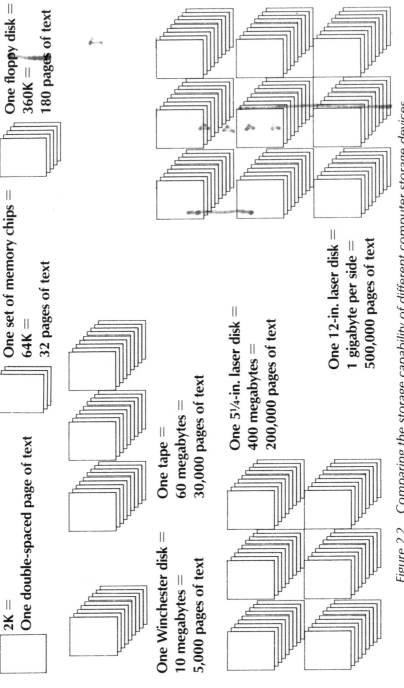

2K =
One double-spaced page of text

One set of memory chips =
64K =
32 pages of text

One floppy disk =
360K =
180 pages of text

One Winchester disk =
10 megabytes =
5,000 pages of text

One tape =
60 megabytes =
30,000 pages of text

One 5¼-in. laser disk =
400 megabytes =
200,000 pages of text

One 12-in. laser disk =
1 gigabyte per side =
500,000 pages of text

Figure 2.2   Comparing the storage capability of different computer storage devices.

With the current IBM PC operating system, the maximum amount of memory that can be addressed without bank switching is 640K. Several memory-expansion techniques allow more than 640K to be used. One technique, supported by Lotus Development's Lotus 1–2–3 and Symphony and Ashton–Tate's Framework, allows up to 4 megabytes of RAM to be addressed and used. The hardware used in this technique, Above Board, is manufactured by Intel. To use these techniques, compatible memory cards and software must be purchased. If you are considering additional memory for this type of expansion, be sure both the hardware expansion cards and the software programs are compatible. Your current software, unless modified by the manufacturer, will not be able to use the additional memory.

Bus Structure

Connecting a number of computing devices together is often necessary. The devices can be chips on a printed circuit board, printed circuit boards in a computer, computers in a local area network (LAN), or huge computers in an international network. One way to connect these devices or units is to connect them to a central location or controller. This strategy, however, limits the access of all devices and limits the number of devices.

Another connection strategy is a *bus,* which connects each device in turn. The electrical wiring in modern homes uses a bus structure to supply electricity. Similarly, the computer chips in a computer are connected by a bus that supplies power and allows the communication of data and instructions. Data and instructions are communicated through the bus in parallel as opposed to serially. *Parallel communication* allows a set of bits to arrive at a component at the same time. *Serial communication* sends one bit of data at a time, and the bits must be translated back to a set of bits.

Just as the selection of a specific CPU dictates how much memory the computer can address, the CPU also dictates the number of bits of data that can be communicated on the computer's bus. For example, the Z–80 processor has an 8-bit bus. That is, 8 bits of information can be communicated to other computer components at a time. More advanced processors, such as the Intel 80286, have 16- or 32-bit buses. These processors can communicate more bits of data in a step.

In most PCs, the bus is extended to include one or more expansion slots. Additional printed circuit boards (called expansion boards) with the proper edge connector can be inserted into a slot. The new board can then obtain power from the computer and have access to the data–instructions bus. These expansion slots allow new applications to be performed without a new or different computer when additional hardware is required. Expansion boards for the IBM PC, for example, provide additional memory and communication ports, perform local area networking, drive various monitors, control data collection, send and receive data from other computers, and perform hundreds of other applications. Expansion slots allow a computer to be customized for a specific application while still being used for other applications.

Ports

A computer communicates to devices outside its own enclosure through devices called *ports*. Communication using ports requires hardware and software. Usually, the hardware is supplied in the form of an expansion board as shown in Figure 2.3. The most primitive software for communication through a port is supplied by the operating system, and specific control is provided by an application program. The two major types of ports are serial and parallel.

Serial ports send and receive data one bit at a time. The best known serial port is a communications port that uses RS–232C protocol to send and receive characters. RS–232C is a communications protocol that defines the electrical characteristics for the transmission and reception of data. Serial ports are also called *asynchronous communication ports*.

Parallel ports are more efficient than serial ports because they can send many bits of data simultaneously. The best known parallel port is the printer port on an IBM PC. This port sends characters to a printer 8 bits at a time. Thus, a complete character is sent at one time.

Keyboard

The keyboard is the main input device for a PC (Figure 2.4). Pressing the keys on the keyboard sends characters to the computer. A typical keyboard is similar to a typewriter. It has keys for the letters of the alphabet and for numbers. In addition, a keyboard has programmable function keys, arrow keys, a numerical key pad, and special function keys. The programmable function keys allow easier user interaction

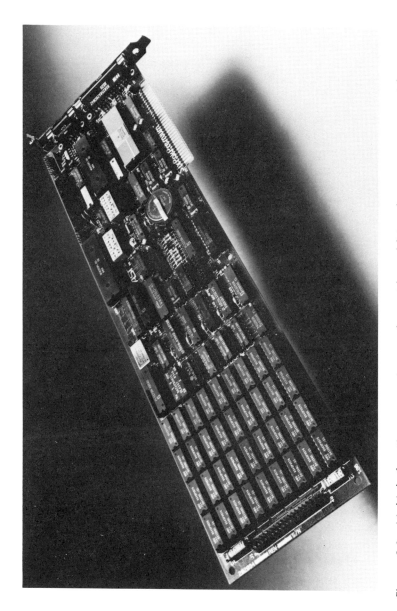

Figure 2.3 Multiple-function expansion card provides additional memory, two serial ports, a parallel printer port, and a clock. (Photo courtesy of AST Research.)

Figure 2.4   The IBM PC/AT keyboard. (Photo courtesy of IBM.)

with a program. Rather than having to type in a specific set of characters, the user of a program can perform a specific task by pressing a function key. The arrow keys allow the user to move the cursor to different positions on the computer screen. The numerical key pad is useful when numerical data must be entered. The special function keys like the Ctrl (control) and Alt (alternate) keys are used like the shift key on a typewriter. To use them, hold them down while you press another key. This double key sequence sends a different character code to the computer. For example, on the IBM PC, when the lower-case c key is pressed, the character with the ASCII value 99 is sent to the computer. Holding down the shift key and pressing the upper-case C key sends the character with the ASCII value 67. Holding down the Ctrl key and pressing c sends the ASCII value 3.

Disk Drives

Disk drives provide read–write storage for programs and data. The data on a disk is not volatile and thus is not lost if power is removed. Two popular types of disk drives are floppy disks and fixed or Winchester disks. Floppy disks on the IBM PC have a capacity of 360K (360 × 1024 bytes) for data or program storage (Figure 2.5), and the PC/AT floppy disks have a capacity of 1.2 megabytes (1.2 × 1,048,516 bytes). Winchester disks have much larger capacities that range from 10 megabytes (10 × 1,048,516 bytes) to 300 megabytes (Figure 2.6).

Disk drive technology is advancing rapidly. With each technological advance, the price of information storage is getting cheaper and is provided in smaller packages. Two new technologies, optical disks and low-cost removable disks for nonvolatile information storage, will become available soon and will become popular on the PC for laboratory applications. Optical disk technology will allow an entire library of information (more than 500 megabytes) to be used in the near future in a briefcase-sized computer using a disk small enough to fit in a pocket. This technology has already been used in consumer

Figure 2.5 Floppy disk drives provide removable data and program storage. (Photo courtesy of IBM.)

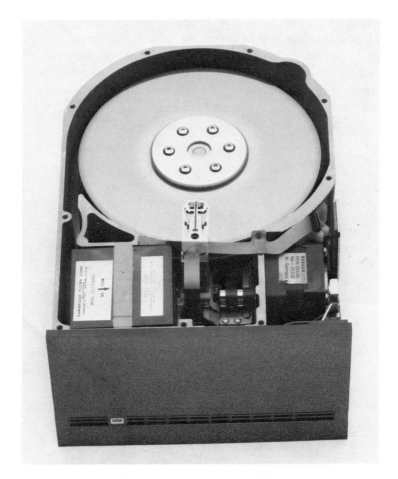

Figure 2.6 Winchester or fixed disk drives provide large storage capacity at attractive prices. (Photo courtesy of IBM.)

products for recording music digitally on compact disks. The first versions of these products are being released, and more will be available. Companies have already made several data bases available on optical disk, including *Chemical Abstracts* and *Grollier's Encyclopedia*. One of the problems with this technology is selecting the data formats to be used. Once this important aspect of optical disk storage is settled, software and hardware vendors will rapidly provide products using this technology.

Lower priced removable disk storage will also soon be available. Currently, disk drives with removable media cost approximately $3000. Prices for this type of drive should be available soon in the

How Disk Drives
Store and Read Data

On magnetic disks—either floppy or hard disks—data is stored one bit of data at a time by using a magnetic head that passes over the metallic coating of the disk. A blank, formatted disk is conceptually similar to metal filings that have been aligned into a single-line pattern by placing a magnet under the paper. When the disk drive's head writes a bit to the disk, the magnetic polarity of a pocket of the filings is changed into one of two configurations representing a 0 or 1. To read this data, the magnetic head again passes over the metallic coating and reads either a 0 or 1 configuration for the specific bit.

The same strategy for data storage is used with an optical disk except different methods are used to create and read the areas on the disk representing one bit. A high-power laser is used to create the minute changes. When the narrow beam of laser light strikes a blank disk, the heat from the laser leaves a tiny pit or bubble, depending on the exact design of the disk drive, on the disk's coating surface. The pits or bubbles can then be read with a lower powered laser. The advantage of the optical disks over magnetic media is the minute area of the disk required to store one bit of data. Normal floppy disks can store 96 circular tracks of data per inch; hard disks can store about 800 tracks per inch. An optical disk can pack 40,000 tracks per inch.

Currently all of the commercial optical disks are read-only; that is, data already prepared on a specific optical disk can only be read. In the near future, advanced optical disk systems will be able to write and read data.

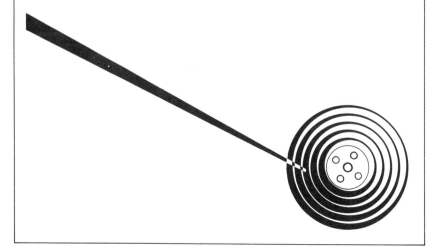

$1000–$1500 range. For this technology to be widely used, it must be priced competitively with the new optical disks and conventional tape storage systems.

Monitor

The *monitor* is the main device used to display entered data and results. The two main types of monitors are text single-color monitors (also called monochrome displays) and graphics monitors, which can produce text and graphics on the same monitor. Monitors are connected to the PC via a special display board that controls what is displayed on the screen. Typical displays and display cards for the IBM PC/AT are shown in Table 2.2. The character clarity of a display depends on the number and density of dots (also called pixels or pels) that can be controlled and displayed on the screen. The more pixels, the better the resolution. Higher resolution screens are more expensive and are supported by less software; therefore, the highest resolution on the graphics screen may not be the best strategy. In most cases, a screen with good resolution (640 × 300 pixels) combined with a plotter will provide the graphics capabilities needed for most applications.

Three standards have emerged for IBM PC graphics: monochrome display with Hercules Graphics Card, RGB monitor with IBM color–graphics card, and IBM Enhanced Graphics Adapter with Enhanced Graphics Monitor. For graphics-intensive applications, the professional color monitor and card provide excellent hardware, but the software must be able to support the card and monitor.

Printer

The *printer* makes permanent copies of information on paper. In most laboratory applications, printing results and reports is the slowest step

Table 2.2 Common Monitors and Cards for the IBM PC, PC/XT, and PC/AT

Monitor	Card	Resolution (pixels)	Colors	Graphics
Monochrome	Monochrome	720 × 348	1	No
Monochrome	Hercules	720 × 348	1	Yes
Medium-color resolution	Color graphics adapter (CGA)	{ 640 × 200 { 320 × 200	1 4	Yes Yes
High-color resolution	Enhanced graphics adapter (EGA)	640 × 350	16	Yes
High-color resolution	Professional graphics adapter (PGA)	640 × 480	256	Yes

in the analysis and reporting of data. Luckily, in the past few months, faster, cost-effective printers have become available from a number of reliable sources. Printers for PCs now come in four main types: dot-matrix, letter-quality, ink-jet, and laser.

The dot-matrix and letter-quality printers use the same impact printing technique—an inked or carbon single-pass ribbon presses against the paper to form an image. Ink-jet printers spray ink in the appropriate patterns to form characters. Laser printers fire a laser at a sensitized surface. The surface that the laser hits is altered so that carbon will stick to that location. A sheet of paper is then pressed against the surface and the carbon is transferred to the paper. The printing process for a laser printer takes only a few seconds per page. Low-cost laser printer technology promises to have many applications in the laboratory.

With a dot-matrix printer, a set of wires in the print head strikes the printer ribbon to form a character. Thus, each character is composed of a set of tiny dots (Figure 2.7). The more dots, the closer the quality to a letter-quality printer. Many of the new dot-matrix printers have the capability to print characters that are almost letter-quality. Although their print quality is normally not as sharp as a letter-quality printer, dot-matrix printers are less expensive and print at much faster rates (100–500 characters per second for dot-matrix vs. 10–60 characters per second on letter quality). These printers can also be used to copy the bit-mapped displays from the IBM PC's graphics

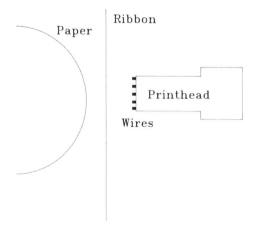

Figure 2.7   Dot-matrix printing is formed by a set of dots created by wires pressing against an inked ribbon.

screen, and with the appropriate software, these printers can produce scientifically useful symbols such as summation, integration, and even some chemical structures.

A *letter-quality printer* is also called a daisy wheel or thimble printer, named after the shape of the actual print mechanism used to print the characters. The letter-quality printer can produce documents that look exactly like typewriter copy. This printer is the ultimate in print quality. Daisy wheel printers use a print mechanism with an interchangeable wheel; each spoke carries a raised character. When the correct character is positioned, a hammer hits the back of the character, pressing it against a ribbon and the paper. Different character fonts (character styles) can be obtained by changing print wheels. Most letter-quality printers can print about 15 characters per second; the more expensive ones print 30–40 characters per second.

Ink-jet printers are interesting printing devices because characters are formed by a spray of ink aimed at the paper (Figure 2.8). This new technology has the advantage that a hammering device is not continually flying against a hard surface; thus, physical wear on the printer is reduced, and almost silent printing results. The character quality depends only on how precisely the ink can be sprayed on the paper. Ink-jet printers have about the same speed and resolution as low-speed dot-matrix printers. This technology is in its infancy, so don't be surprised to see outstanding print quality and speed from these devices in the near future. But this technology must compete with laser printers to be successful.

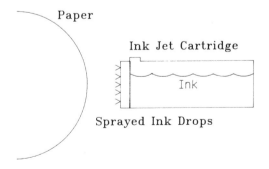

Figure 2.8   Ink-jet printers form letters by spraying ink at specific locations on the paper.

Laser printing is a new technology for PC printers (Figure 2.9). This type of printer has been used extensively on mainframe and minicomputers for more than ten years. The technology is not new, but certainly the prices for a laser printer are within reach of individuals. The outstanding features of this printer are its speed and printing versatility. The speed of a laser printer is not measured in characters per second but in pages per minute. Current low-cost models print three to eight pages per minute. These printers can also create different fonts, letter sizes, and print graphics at these speeds. These printers are excellent for laboratories that require presentation-quality output. The speed is fast enough to allow many users to share on a network. Laser printers promise to provide presentation-quality reports with near-typeset-quality print, including multiple fonts mixed with crisp, near-plotter-quality graphics. The best current technology provides approximately 300 dots per inch; nearly 1000 dots per inch will be available in the future for the same price range.

Because of the improved print resolution and low cost of laser printers, many laboratories that currently print their reports on specific forms should consider printing both the forms and the results at the same time. Using a laser printer and the appropriate software, the PC could fill out the form, and then the form and results could be printed together.

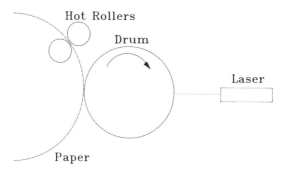

Figure 2.9 Laser printers are a combination of printer and office copier technologies. Lasers write the image on a copier drum. The image is then used to attract toner that is trans- ferred to paper, just as in xerographic copying. The high resolution of this process allows both text and graphics to be mixed on the same page.

The effect of the PC revolution is easily measured in lower prices and higher performance when printers and plotters available today are compared with those prices for equivalent printers and plotters only three years ago (Table 2.3).

Plotter

A plotter is needed to obtain high-quality graphic output on paper or transparencies from a computer. *Plotters* are mechanical devices that can hold a pen. The pen is then moved across the paper to draw graphs or characters. The pens can be colored to create multicolored graphs. Multipen plotters can change pen colors as part of their repertoire.

Two main types of plotting styles exist. One plotter type allows you to lay the paper or transparency on a flat plotting surface. The plotter can then control a pen in two dimensions over the paper (Figure 2.10). The other plotter type moves the pen in one plane and moves the paper to obtain the second dimension (Figure 2.11).

Plotters get their instructions on what and where to plot as a series of instructions from the computer. Proper hardware and software are needed to operate a plotter. The hardware consists of a communication interface and cable. The software must send the right sequence of characters to the plotter to make the drawing. Most plotters have set codes that are translated into pen movements by instructions programmed in the plotter. The plotter commands from the computer are given in a command language. No official standard for plotter language is recognized; however, Hewlett–Packard Graphics Language (HPGL) has become a de facto standard because of the large number of plotters that have implemented the language and the large number of software packages that support plotters that use HPGL. The Houston Instruments Graphics Command Set is a distant second in the graphics language race. The International Standards

Table 2.3   Reduced Price of Printers

Printer Type	1983 Price ($)	1986 Price ($)	Change (%)
Matrix	995	295	–70
Daisy wheel	2,495	895	–64
Ink jet	NA	495	—
Laser	19,995	2,995	–85
Six-pen plotter	8,595	995	–88

Figure 2.10 Flatbed plotter, where the pen can move in the X and the Y directions. (Photo courtesy of Hewlett–Packard.)

Figure 2.11   Paper movement plotter. The paper is moved by rollers in the X direction while the pen is moved in the Y direction. This method of plotting requires fewer mechanical parts and is as precise as a flatbed plotter. (Photo courtesy of Hewlett–Packard.)

Organization (ISO) is working toward a standard plotter command set; therefore, this area will be less of a problem in the future.

Bar Code Reader

Another input device that is quickly gaining popularity in scientific applications is the *bar code reader*, shown in Figure 2.12. Bar codes appear on most grocery packages and many consumer products. *Bar codes* are identification numbers that are encoded as a series of narrow and wide lines. The lines can be scanned with a bar code reader, and the numerical or alphabetic characters can be read into a computer. Bar codes and bar code scanners are used extensively in retail sales. The bar code on a package can be scanned, the package identified, and its price and description printed on a receipt in a fraction of the time required for the clerk to enter the information manually. Inventory bar codes are also great labor savers for counting.

Bar code applications in the laboratory do not differ from those in a retail store. Bar codes can be used to identify everything from samples and instruments to reagents and laboratory personnel. Bar coding can easily be integrated into current software programs

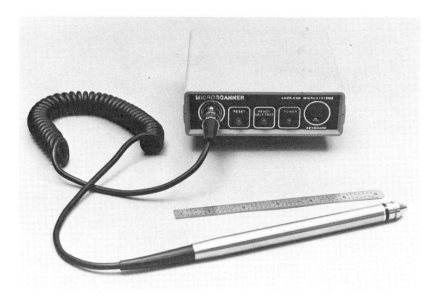

Figure 2.12 Bar code reader–scanner wand and a bar code. This model plugs into a standard PC keyboard. (Photo courtesy of American Microsystems.)

because most bar code readers for the IBM PC have been engineered to reside on the same data entry line as the keyboard. The program cannot distinguish between manual data entry or bar code reading.

Bar codes can be printed with dot-matrix, ink-jet, or laser printers (normally not with letter-quality printers) and the appropriate software. A number of bar codes are currently used for various applications. Be sure the bar code, bar code scanner, and bar code printing software are compatible.

Mouse

A *mouse* is a pointing device that allows a cursor to be manually operated on the screen quickly and easily (Figure 2.13). These devices are so named because a wire cord, or tail, connects them to the computer. Mice have become popular with the introduction of full-screen, graphics-oriented program interfaces. The mouse allows you to move to an icon on the screen and select its operation by pressing one of the buttons on the mouse's "back".

Figure 2.13 The Microsoft mouse has two buttons and is interfaced to a PC either via RS–232C or directly with a plug-in interface card.

Like other hardware products, the software selected must be able to recognize that the mouse is being used and recognize the controls. Luckily, most software packages that use the PC cursor keys for cursor movement can use a mouse after a small patch program is installed to redirect the mouse's movements as cursor keys are pressed.

The two broad classes of mouse designs are mechanical and optical. A mechanical mouse uses a rolling ball that rotates in response to friction as the mouse is moved over a desk surface. As the ball rotates, internal parts sense movement, count the number of ball rotations, and send this information to the computer. Optical mice move over a special reflective pad. A beam of light emitted by the mouse is reflected by evenly spaced lines embedded in the pad. A sensor inside the mouse can then track the distance and direction the mouse has traveled and send this information to the computer.

Mice come with one, two, or three buttons on their backs. How these buttons are used depends on the software. The one-button mouse is used only with the Apple Macintosh; the two- and three-button models are used on IBM PCs. The Microsoft mouse with two buttons has become the de facto industry standard. Mice communicate with the PC either through an RS–232C serial port or through an expansion interface card plugged directly into an expansion slot.

Other, less popular pointing devices available for use on the PC include joy sticks, light pens, touch pads, track balls, and touch screens. All of these devices allow the user to select from either menus or graphics displayed on the screen as part of program input. For some laboratory applications, these alternative input devices could be useful.

Math Coprocessors

Math coprocessors are optional chips that are easily inserted into open chip slots on the mother board of an IBM PC, PC/AT, and clones. These devices add additional mathematical or arithmetic instructions to the instruction set of the main microprocessor. Until recently, the software necessary to use these devices had to be written specifically for a PC with a math coprocessor and would not work without the coprocessor present. Now, some programming languages can sense the presence of the coprocessor and use it, or if no coprocessor is present, the programs will still operate.

If the software can use the math coprocessor, computational speeds can be increased by factors of 4 to 20. The actual decrease in computing time that results from using software and the math coprocessor depends on the application. The coprocessor speeds only mathematical instructions, not disk access times or text-processing times. Thus, unless the application is slowed by mathematical computing, the use of a coprocessor may not provide significant improvement.

The math coprocessor used on PCs is the 8087 to complement the 8088 microprocessor. The PC/AT and AT clones use the 80287 coprocessor to complement the 80286 microprocessor.

Products Mentioned in This Chapter

Microsoft Mouse ($195 serial version, $175 bus version), Microsoft Corp., 10700 Northup Way, Box 97200, Bellevue, WA 98009. (800) 426–9400

PC Mouse ($195), Mouse Systems Corp., 2336H Walsh Ave., Santa Clara, CA 95051. (408) 988–0211

Bar code scanner ($495), American Microsystems, P.O. Box 830551, Richardson, TX 75080. (817) 834–9659

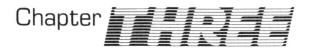

Operating System Software

oftware programs are generally divided into two classes: system software and application software. The majority of this book is concerned with *application software,* which is made up of programs that perform specific applications such as word processing, spreadsheet analysis, data management, graphics, or chromatographic data analysis. This chapter covers the other kind of software required to operate a computer, systems programs. *Systems programs* are software tools that make the computer easier to use regardless of the application program. In current PCs, the functions of the operating system are taken for granted by most users. One cannot imagine computers without these functions, though just a few years ago when PCs were in their infancy, operating system functions did not exist. These programs are so hardware-dependent that they are sometimes thought of as part of the computer. In fact, part of the IBM PC's operating system is permanently stored in ROM. Four types of systems software will be discussed in this chapter: operating systems, application environments, hardware enhancers, and programming languages.

Operating Systems

The *operating system,* as its name implies, controls nearly every aspect of the basic operation of the computer. The operating system provides two major interfaces: one between the user and the PC, and another between the application program being executed and the various pieces of hardware available in the PC. This section describes the functions available from the operating system.

1000–4/87/0045$07.50/1   © 1987 American Chemical Society

The User Interface

The moment the computer is turned on, the operating system takes control. The first function the operating system is programmed to do is to test parts of the PC hardware to see if the hardware is functioning correctly. These tests on available memory, interfaces, and other system components are performed during the long delay experienced after power is applied to the PC. If all checked hardware is functioning correctly, the remaining portion of the operating system is loaded from disk.

If a file on the current default disk has the name AUTOEXEC.BAT, the commands contained in the file are executed. This file allows a computer's operation to be customized so the user has control as soon as the computer is switched on. If no AUTOEXEC.BAT file exists, then the user will be prompted for the date, time, and operating system prompt, which is the letter of the current default disk drive followed by an angle bracket, **A**>. The computer, through the operating system program, is asking which application program or operating system function to perform next. The response to this prompt can be the execution of a program, which is done by entering the name of an application program, or the performance of one of a number of operating system functions. Some common PC–DOS commands are shown in Table 3.1. All of the PC–DOS commands listed are called *reserve words* because an application program cannot have the same name. Most of the commands listed control and communicate with the main hardware components that include the disk drive, keyboard, monitor, and printer. These command functions are the most basic ones needed to operate a computer. They provide information on the size of disk files, the date and time, the disk files that were last created, and the space left on disks. These commands perform specific tasks like copying, renaming, and deleting files. The operating system programs also load application programs and begin their execution.

When the command to load and execute a program is given, the operating system will first look for the program on disk. After finding the program, the operating system gets the starting disk storage location and loads the program into available memory. Then the control of the computer is turned over to the application program. From that point until the program is exited, only the application program is dealt with. When the application program is no longer in

Table 3.1   Some PC–DOS Command Functions

PC–DOS Command	Function
PROGRAM NAME	Load the program into memory and begin execution
CD	Change directory to a different subdirectory[a]
CHKDSK*	Check the size and number of files on a disk
COMP	Compare the contents of two files
COPY	Make another copy of a file or send the contents of a file to another device[b]
DATE	Set system date
DIR	Display a directory of the disk
DISKCOMP	Compare the contents of two disks
DISKCOPY	Copy an entire disk to another
ERASE or DEL	Erase or delete a file from the disk
FORMAT*	Prepare a disk for data
MD	Make a new subdirectory
MODE*	Change the current mode of operation on a device
RD	Remove a subdirectory from the disk
RENAME	Change the name of a file on the disk
SYS	Copy the operating system to a disk
TIME*	Set the system time
TYPE	Display the contents of a file

[a]A *directory* is used to store files in a disk or diskette.
[b]A *device* is a physical entity like the printer, communications port, monitor, or disk file.
*All of the commands except those marked with an asterisk are executed as part of the operating system. Those marked with an asterisk are programs that are loaded when the command is executed. These programs must be on the current default disk to be executed.

use, control will be returned to the operating system for another command.

Command Line Editing

The operating system also supports some useful command line editing features that use the function keys and other keys on the keyboard. For example, if the wrong letter on the keyboard is typed, that entry can be deleted by pressing the Del (delete) key near the numerical key pad. Function keys 1 (F1) and 3 (F3) are useful if the same or a similar command has to be re-entered. By pressing F3, the entire command line just entered will be retyped. An example of this command use is if the CHKDSK command is used to check the contents of both the A: and B: drives. In PC–DOS or MS–DOS, disk drives are designated by using a letter of the alphabet followed by a colon. If you have two floppy disks in your computer system, the drives are called the A: and B: disk drives. The CHKDSK command prints the number and size of files on each disk. You can type **A>CHKDSK A:** to get the status of the A: drive. To do the same on the B: drive, simply press F3 and the previous command line will appear as **A>CHKDSK A:**. Before pressing the Enter key, press the Del key twice to remove the **A:**; then replace it with **B:** and press Enter. Using the F3 key in this way saves keystrokes.

Pressing F1 types one letter of the previously used command line at a time. Again, this typing aid saves repeated keying of specific characters. As an example, perhaps you want to type the contents of three files with the names ABCD1.TXT, ABCD2.TXT, and ABCD3.TXT. You can type the first command as **A>TYPE ABCD1.TXT.**

Now, rather than having to retype the command, hold down the F1 key and watch the characters appear on the command line as **A>ABCD**. After ABCD has appeared, release the F1 key and press the number 2 to replace the number 1. This should appear as **A>ABCD2**. Now, press F3 to fill in the rest of the characters followed by the Enter key. The same keystroke pattern can be followed to type the last file. This strategy of keystroke conservation will become second nature after a while.

You can also use the Ins (insert) key to insert new characters to the command line. Using a similar example, suppose file ABCD3.TXT has just been typed, and now the file ABCD123.TXT needs to be typed. Hold down the F1 key while **ABCD** is typed. Then press the Ins key once, and then type **12**. The result should be **A>ABCD12.TXT**. The line can be completed by pressing the F3 key.

Function Key Setting

With DOS 3.1 and the previous examples, you can reprogram the function keys to execute a series of DOS commands. This programming allows optimum efficiency for your computer. If you are typing the same keystrokes over and over again, then you can reprogram a function key to execute the keystrokes. Suppose two different directories are needed for chromatographic software and word processing software (see the discussion on disk file directories in this chapter if you are unfamiliar with hierarchial directories). Every time you want to go to the word processing (WORD) directory from the chromatography (CHROM) directory, enter **CD**, then type **WORD**. With the command **ASSIGN**, those commands can be performed by pressing F7. Similarly, the F8 key can be reprogrammed to go to the CHROM directory and bring up the chromatography software main menu. Use this feature with caution. Reprogramming function keys can cause some application programs that use the same function keys to perform oddly or incorrectly.

Interface between Application Programs and the Computer

The other interface the operating system provides is between an application program and the computer's hardware. The main hardware component that the operating system is concerned with is the disk. The disk operating system (DOS) keeps track of the physical location of all data and programs on a disk. Figure 3.1 shows a directory of disk files on a disk. The directory is displayed when the **dir** command is entered.

If a large capacity disk is available, the operating system allows easy division of the disk into logical units, just like different drawers in a filing cabinet. These divisions are made using hierarchal, or tree, directories. An example of these directories is shown in Figure 3.2. These directories allow hundreds of files to be on a single disk and allow files to be placed logically together in the same subdirectory. Operating systems without hierarchial directories are very cumbersome to use with large, high-capacity disk drives.

The operating system also commands the connection and controls input and output devices such as the keyboard, monitor, disk files, printers, or communication ports through programmable devices within an application program. The programs that perform these

File	File Type	Characters	Date	Time
GPCCAL	WKS	3072	7-29-85	9:34p
SIMDIS	WKS	2304	7-29-85	9:52p
QACHART	WKS	3584	6-16-86	5:20p
BEVERAGE	WKS	2432	7-30-85	3:58p
AREAPER	WKS	2560	11-16-85	12:19p
SCIPIE	WKS	1536	11-24-85	5:46p
EXTSING	WKS	4352	11-24-85	6:28p
MULTIGRP	WKS	4608	1-26-86	1:56p
ICPDATA	WKS	4096	6-16-86	4:15p
BARS	WKS	4480	6-16-86	5:26p
10 File(s)			38912 bytes free	

Figure 3.1   Directory of files on a disk.

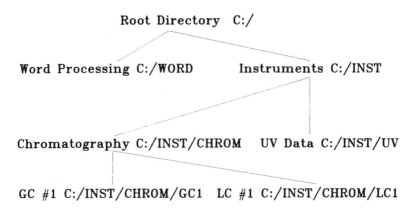

Figure 3.2   *Program and data files can be placed in subdirectories for easier access. Similar files and programs can reside in the same subdirectory.*

invaluable tasks are called *device driver programs,* or *drivers* for short. Specific drivers must be written for each type of device. This function of the operating system greatly benefits the developers of application programs. If this facility were not present in PC–DOS, then each application program would have to include a code to perform such mundane tasks as reading keystrokes, placing the pressed characters on the monitor, or sending characters to a printer. Because these functions are preprogrammed and readily available through simple programming, the time required to develop an application is substantially decreased. Application programmers have access to all computer hardware devices through easy-to-program device interfaces. The operating system allows control over which devices will provide input and output for the application programs (Figure 3.3). This control is very valuable because no modification of the application program is required to send, for example, printed results through the communications line to a letter-quality printer rather than through the normal printer port where a dot-matrix printer is connected.

Other Operating Systems Available for the IBM PC

PC–DOS is only one of many operating systems that has been developed for the IBM PC. Many more operating systems will run on the IBM PC, but they are not as popular because these systems do not have as many application programs written for them. Just as the

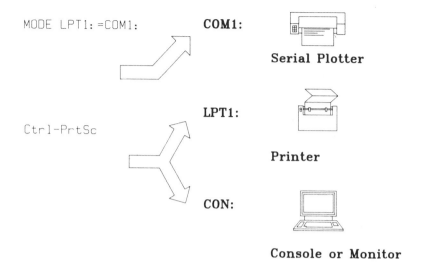

```
MODE LPT1:=COM1:
```
COM1:

Serial Plotter

LPT1:

```
Ctrl-PrtSc
```

Printer

CON:

Console or Monitor

Figure 3.3 The operating system allows you to connect and disconnect various devices. To send output to COM1: rather than LPT1:, use the MODE command. The keystroke combination Ctrl and PrtSc sends all output that normally goes to the computer monitor to go also to the LPT1: printer port.

engine in a car is designed for a specific model, a given application program must be written to run under a specific operating system. Thus, when you select an operating system, you also define and limit the spectrum of application programs. Here is a quick review of some other operating systems that operate on the IBM PC. Each of these operating systems has some specific performance features that are attractive; however, none of these systems has the application program support found with PC–DOS on IBM PCs or MS–DOS on PC compatibles.

Concurrent, Multitasking, and Multiuser Operating Systems. The major feature that the alternative operating systems provide is the ability to execute more than one application or task. PC–DOS is a single-user, single-task operating system: A person can execute one application program on the computer. This mode of operation is simple and powerful. It is simple because the operating system needs to control only one application program, and it is powerful because the single user has the full resources of the computer. The operating system can be much simpler, and application programs will have very few restrictions; these programs are thus easier to develop and maintain.

Historically, when computers were very expensive, a computing resource had to be shared to make computer use economically feasible. Thus, mainframe and minicomputers normally have multiuser, multitasking operating systems so the computing resource can be shared with more people. By their very nature, these operating systems are more complex and difficult to use. Application programs written for these systems have restrictions on operation, and these restrictions make program development longer and more expensive. Mainly for economic reasons, multiuser, multitasking operating systems have been developed for the IBM PC. Most have been used for data entry through lower cost terminals or for applications where many users need to communicate information to each other. Now, however, almost all of the saving is done because the prices of a stand-alone PC and of a data entry terminal are within $500. The introduction and wide use of low-cost local area networks (LANs) (see Chapter 9) has also removed the economic incentives of multiple users on a single computer. I would not consider using a PC multiuser system today because of the complexity of operation and the lack of application programs and hardware support. Purchase of a second or third PC to perform other tasks is more cost-effective than use of a multiuser system.

For scientists, a new combination of features, single-user multitasking, is very attractive. These systems are also called concurrent, which means that two to four programs can be executed seemingly at the same time. This process is attractive to scientists because two or three of the programs can perform data acquisition, control, or both while another application program is run interactively. The interactive program could be used to review previously collected data or write a report using a word processor. Concurrent operation is available now using PC–DOS or MS–DOS when one of the application environment programs like Topview or Microsoft Windows is used. This process is a very attractive way to increase productivity without sacrificing the valuable base of application programs available for PCs that use PC–DOS. For laboratories that need to have multiple users sharing information, a local area network is the most cost-effective solution.

CP/M–86 and Concurrent PC–DOS. The original version of Digital Research's CP/M–80 was available on hundreds of different Intel 8080 and Zilog Z–80-based 8-bit computers. This version is still the de facto standard operating system for 8-bit computers. Although many of the 8-bit application programs were transported to the IBM PC, most were written to run on PC–DOS rather than CP/M–86 because PC–DOS, including the Microsoft BASIC interpreter, was available for $60, and CP/M–86 was more expensive. The number of application programs for CP/M–86 has never come close in popularity to the number of Z–80-based and Apple II-based (using a Z–80 card) programs that run on CP/M–80. Concurrent CP/M–86 was the first multitasking operating system on the IBM PC. Digital Research followed this product with Concurrent PC–DOS, which runs up to four PC–DOS programs. The first version could run only PC–DOS revision 1.1 programs. Now a later revision can run PC–DOS 2.0 programs.

Because a de facto standard operating system is available and widely used, there is no reason to run an operating system that mimics the original. Concurrent PC–DOS does not support the IBM PC network or other networks. Some application programs written for PC–DOS may not operate together in the multitasking mode, and the only way to discover if these programs will run or not is to try them out. This situation requires time for testing program compatibility. Your time is probably better spent learning more about application programs.

UNIX on the IBM PC. PC/IX and XENIX are derived from UNIX, an operating system developed by Bell Laboratories in the early 1970s and widely used at universities. PC/IX (for interactive executive) was ported to the 8088 microprocessor from AT&T System III sources by Interactive Systems. Although UNIX is normally a multiuser, multitasking operating system, this implementation of UNIX is single-user and multitasking.

UNIX or any other single-processor, multitasking operating system is not a good choice for real-time laboratory applications where timing is critical. Only when the real-time portion is communicating with an intelligent device or a device designed with the proper communications buffer and protocols can real-time applications be performed without error. A much better strategy, both for economic reasons and for fast implementation, is to dedicate a single-tasking PC to the real-time application.

XENIX is a multitasking, multiuser operating system for the IBM PC developed by Microsoft.

UNIX has many operating system features that are handy for experienced UNIX programmers. UNIX is still essentially a programmer's operating system with very few application programs compared with the number available on PC–DOS.

Application Environments

Application environments could easily be mistaken for part of the operating system. *Application environments* are actually application programs that run on PC–DOS but provide a more advanced user interface and other interesting features, including concurrent operation, coresident programs, and graphics compatibility. The only major drawback to these environments is that to use all facilities, an application program must be written to include a code that interacts with the application environment.

Topview from IBM

IBM markets Topview as a multitasking extension to PC–DOS. Topview does provide multitasking, but not all PC–DOS application programs are compatible and run when Topview is executing. Topview implements multitasking by assigning each application a time slice of the central processor and executing each program in round-robin fashion. One application, which is called the foreground task, interacts with the keyboard and computer screen. Other active applications, called background tasks, execute at a slower speed and cannot interact with the keyboard or screen. A background task can be quickly changed to a foreground task by pressing a mouse button or combined keystrokes. Tasks can be displayed in overlapping sections of the screen.

Software developers have been reluctant to allocate resources to the development of programs or even conversions of current programs that run in the Topview-specific mode of Topview. Because the full potential of Topview will only be realized when a large number of Topview-specific applications are available, users have delayed procuring Topview until more applications appear. As long as developers and users wait for each other to lead the way, acceptance of Topview will be slow.

For a program to operate under Topview, the program must meet some specific conditions. Programs cannot use absolute memory addresses or be loaded at a specific address. Programs cannot directly access the video memory or the keyboard buffer. Batch files are not supported, nor is graphics mode.

Topview currently has several notable deficiencies besides not having enough compatible programs to run. Topview is not compatible with the IBM PC network because it does not support any of the features of PC–DOS 3.X beyond those of PC–DOS 2.X. Topview requires much larger memory resources than most current PCs contain. Assuming 256K memory, only 80K remains for application programs after Topview is loaded. Even with the maximum 640K memory, only one 256K application program can be run. Operating speed is also disappointing. For example, when only one active task is running, a simple BASIC program requires almost twice the amount of time to execute.

Microsoft Windows

Microsoft Windows is an application environment that loads after DOS and takes control of the machine. Through Windows, two or more application programs can be run, and within some strict limitation, information can be moved between the two applications. Windows is a graphics-oriented interface, first popularized on the Apple Macintosh. The entire screen becomes the command line. Therefore, rather than typing in commands, a mouse or the cursor keys can be used to move to an icon symbol for an application. The application is executed when the mouse button is pressed. DOS utilities are also represented by icons.

Both Windows and GEM (described in the next section) implement a *device-independent graphics interface*; programs that are developed using this facility should run on other completely different hardware environments when the application environment is ported (connected) to other computers. For software manufacturers, this interface is a positive aspect of these operating systems. Of course, the only way these environments will become widely used is if software is written that fully uses them. Lotus Development has announced that its future products will be written to be Microsoft Windows-compatible. Windows displays its operating programs in tiled windows on the screen, as shown in Figures 3.4 and 3.5. Notice the windows cannot overlap.

GEM from Digital Research

Digital Research has developed GEM as a visually oriented user interface comparable with the user interface on the Apple Macintosh. GEM is a single-tasking environment that provides support for display

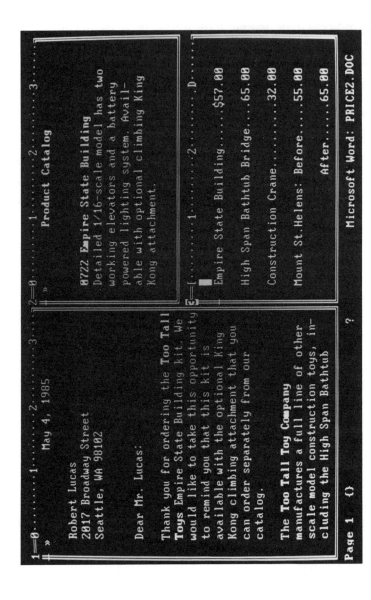

Figure 3.4   Microsoft Windows with the word processor in action. (Photo courtesy of Microsoft.)

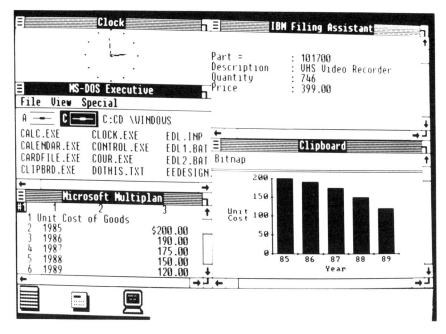

Figure 3.5 *Windows includes a number of desk-top utility programs such as an appointment calendar and calculator. (Photo courtesy of Microsoft.)*

windows, pull-down menus, and pointing devices. GEM is built around graphics that use a desk-top metaphor. An analogy is drawn between computer resources and objects normally found on a desk top. The desk-top objects are represented on the display screen as icons. For example, disk directories are represented by file folder icons, and text files are document icons. Operations are performed on icons rather than typing commands. For example, to delete a file, the selected document icon is placed into a trash can icon. Many novice users will find this graphically oriented user interface easier to learn and use. Digital Research also offers application developers the opportunity to create GEM-specific applications that would include, within the application, GEM-like interaction.

All existing PC–DOS application programs can run under GEM. If the application does not use any of GEM's functions, GEM allows the application to run as usual without any performance degradation. An example of the GEM interface is shown in Figure 3.6.

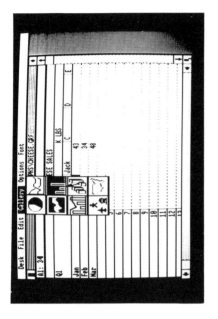

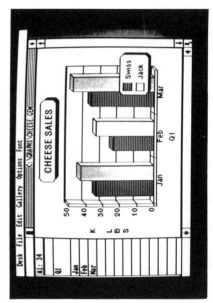

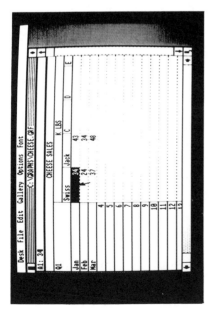

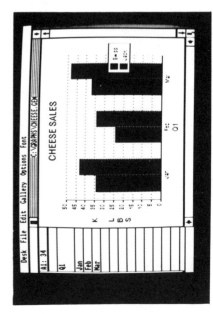

Figure 3.6   GEM graphics interface. (Courtesy of Digital Research.)

Hardware Enhancers

Many unique programs have been developed to enhance the PC's hardware. Most of these programs are great productivity tools because all save time and effort. Some programs are ingenious problem solvers and reflect the spirit of personal computing, which involves getting more done with the same resources.

The Norton Utilities

The Norton Utilities is a disk full of utility programs that perform special tasks you sometimes need. The most famous task is recovering files that are accidently erased or deleted. Sometimes an application program will alter or maim a disk file. With the UnErase program, a file can be recovered as long as the data from any new file has not been written over the old data. The Norton Utilities also includes programs for other disk management uses such as listing disk directories, finding specific files in any directory, or searching disk files for a specific set of words. These programs are highly recommended for piece of mind and a better understanding of the operation of your computer.

Hot Key or Memory Resident Programs

Hot key programs are programs that are loaded into the memory of a PC and wait there until a moment of need arises. These programs then appear on the monitor with the touch of a key, thus the name hot key. While the hot key program is running, the application program that was running is still in memory and can continue its operation with one keystroke. Hot key programs perform many different applications and are available from a host of vendors.

The best example of this type of program is Sidekick, available from Broland International. Sidekick can be loaded as a part of the AUTOEXEC.BAT file, so entering the keystrokes to load the program is not necessary. Sidekick begins operation whenever the keys Ctrl and Alt are pressed at the same time. A menu of functions (Figure 3.7) will appear on the screen. Functions in Sidekick include an on-screen calculator, notepad with text editor, appointment calendar, and telephone directory and dialer if a modem is installed. A *modem* is a hardware device that interfaces telephone lines and hardware equipment. For programmers, an ASCII character conversion table is included in Sidekick.

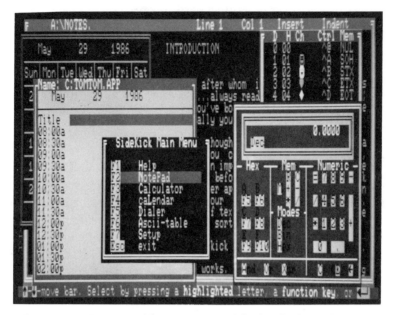

*Figure 3.7   A menu of functions in Sidekick, which is a hot key
program. (Photo courtesy of Borland International.)*

The value of these programs is the immediate availability. If you are
in the middle of editing a report and you need to make a quick
calculation, then you can simply press the Ctrl and Alt keys simultaneously, select the calculator, make the calculation, note the answer,
then press Esc, and you will be returned to where you left off.
Similarly, if you are using a spreadsheet program and someone calls
to ask if he or she can meet with you next Wednesday, you can press
Ctrl and Alt, select the calendar function to see if next Wednesday is
open, answer his or her question, press the Esc key, and you are back
to your spreadsheet application. These programs make a PC fun and
functional.

The notepad is particularly useful. The notepad uses a command
set almost identical to WordStar and does many more operations that
cannot be done with a normal word processor. For example, sections
of a document and even sections of a line can be cut and pasted from
one document to another. Data can be read from the current display
and entered into the notepad. This feature can be used to transfer
text from one program to another.

Suppose you are running a program like Lotus 1–2–3, and someone
calls on the phone and asks about a letter you just wrote or the

amount of a sales order. Normally, you would have to save your work, exit Lotus, load your word processor, and load the file you are interested in. These steps could take one or two minutes. To get back to where you left off, you must reload Lotus and your work. With Sidekick, this whole process can be shortened. Press Ctrl and Alt, select the notepad, read from the letter or sales order, get the answer, and press Esc to go where you left off. Sidekick and programs like it are real time savers.

Electronic or RAM Disks

If more random access memory (RAM) is available than the application programs require, then part of the additional memory can be used as an electronic or RAM disk. These programs allow a portion of RAM to emulate a disk drive. These programs completely simulate the function of a disk except at a much higher speed. During execution, these programs speed any application that uses disk access to read or write data. Because most programs fall into this category, these programs can save execution time.

These programs are usually supplied with a multifunction expansion board. Electronic or RAM disks are easy to use and, like the hot key programs, the load commands can be inserted in the AUTO-EXEC.BAT file. This ability makes their use completely invisible to the user. The electronic disk program divides the PC's memory into two regions. One region is for the operating system and application program. The second region is set aside to emulate a disk drive. The emulated disk is assigned a disk letter, just like a normal disk drive. When an attempt is made to read or write data to the emulated disk, the instructions are intercepted and data transfer takes place between the two sections of memory rather than between a real mechanical disk and memory. Because data is transferred from memory to memory rather than from memory to mechanical disk, transfer times for RAM disks are two to three times faster than those for floppy disks. These transfer times are similar to those for hard disks.

Print Spoolers

Spool is an acronym for simultaneous peripheral operation on line. The term was coined from the operation of large computers in which one system printer can serve many users. With so many users, the system printer could easily be busy printing a job for another user when you need to print. You can still issue the printing commands, but rather than immediately printing the contents of your job, the computer system temporarily stores on disk or in memory the information to be printed. The storage area where your job is kept is called a *buffer* or *print queue*. When the current print job is complete, the next job in the buffer is printed.

With a single-user PC, the need for a print spooler seems remote, although a print spooler is a very valuable tool for increasing the productivity of a system. A print spooler allows work to be done on a PC while a printing job is being printed. Printers are one of the biggest bottlenecks in a system; several minutes can elapse while a printer prints a document. A print spooler will accept all the characters to be printed as if the printer is printing at a fantastic speed, thus the application program is freed from its printing task. The application can be continued while the print spooler sends the characters to the printer at the normal rate.

The two types of print spoolers are hardware and software. Hardware print spoolers are boxes containing RAM. These print spoolers are placed between the computer and the printer. Hardware spoolers come in sizes up to 512K memory. Software spoolers allow part of your RAM to act as a spooler. This software is also available from the companies that manufacture multifunction expansion cards.

The most popular and useful of these operating system enhancement programs usually are integrated into future versions of the operating system. For example, a new print command performs spooled printing as a separate task in PC–DOS 3.0 and later PC–DOS versions. This print command allows you to print while you run another application program. Both tasks run slowly because the computer's processing resources are shared.

While a software spooler is in use, two programs are being executed. The spooler program "steals" computer time from the other application that is running. If a real-time application program is running in which timing is critical, an active spooler program can cause the real-time program to operate erroneously and lead to loss of control, loss of data, or the overflow of data communication buffers.

Programming Languages

Even with the thousands of application programs that do word processing, chromatographic data handling, and data management, you may still have an application that no prepackaged commercial program can perform. If you are like most PC users, you will always have one more thing you would like to do with your computer. This situation occurs when software manufacturers simply cannot keep up with the demands, and you will have to write the program to do the job. This section will introduce you to six programming languages that have been popular with scientific PC users.

Each computer programming language has its historical place and purpose. All programming languages have strong proponents who are willing to challenge the use of any other language on either technical or purely emotional grounds. Computers existed well before the first programming languages. The first computers were programmed by setting switches and wiring "bread boards" that connected the instructions and locations in the computer. When memory was invented, the switch settings could be stored inside the computer memory and then executed. But the program steps had to be entered first in binary code, then later in hexadecimal and octal codes. Finally, computer programming emerged as the binary code program steps were generated from instruction mnemonics through an *assembler*. The assembler took each mnemonic and assembled one computer instruction.

Higher level languages allowed the programmer to concentrate on solving the problem rather then how to instruct the computer to do it. The resulting programming languages required more time to execute than those written on the assembler but could be developed much faster (Figure 3.8). The first of these languages was FORTRAN (formula translation), designed to solve mathematical and scientific problems. For business applications, COBOL (common business-oriented language) was designed to perform accounting and record keeping.

Compiled Languages

Both FORTRAN and COBOL are known as compiled languages. *Compilers* transform the source code entered by the programmer into machine or object code. Source code is the program written in the high-level language. Object code or machine code is instruction that the central processing unit (CPU) can execute. After entry, the machine code is then linked with one or more other machine code modules to form the executable program. If the resulting program does not operate correctly, the original source code must be changed

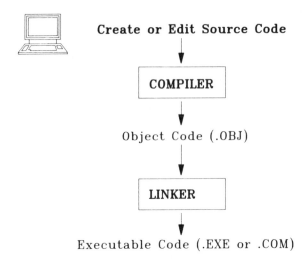

Figure 3.8   Steps involved in writing a compiled program.

and the compiling process repeated. Current, popular compiled languages are Pascal, Compiled BASIC, and C.

Interpretive Languages

Compilers are good at converting source code into machine code but are not good at conserving a programmer's time during program development. Particularly when you are first learning to program, a compiler becomes another program to be mastered before you can feel comfortable with your computer.

With students and computer novices in mind, two Dartmouth College professors, John Kemeny and Thomas Kurtz, developed an alternative to compiled languages, which they called an *interpretive language*. Thus, the language named BASIC (beginner's all-purpose symbolic instruction code) was born.

By their very nature, interpretive languages are slower than compiled ones. An interpreter reads each line of source code, composes the machine code for the instruction, and executes the instruction. But the loss of execution speed is compensated for by the interpreter's ability to enable the user to interactively enter a program, immediately execute it, and see the results without stopping to compile and link the program. For small, quickly needed programs of less than 1000 program lines and for quick learning of programming skills, interpretive languages are very attractive.

Threaded Interpretive Languages

A third category of programming languages is called *threaded interpretive*. The most prominent language in this category is the language FORTH. Programs written in FORTH have the advantage of fast execution, and the language is easily extended to fit the application. FORTH allows the programmer to define new instructions based on other instructions or utilities. The vocabulary of the language can thus grow, depending on the application. Because FORTH executes rapidly, it has been widely used in real-time applications. A number of versions of FORTH are available on the IBM PC and PC clones.

Structured Programming

Computer programming has evolved from hardware toggle switching to its present form by continuously changing and improving. One of the more recent trends in writing programs has been the concept of structured programming. *Structured programming* results in programs that are simple statements of the problem and solution. Thus, the programs are easier to develop and maintain. Although structuring of programs can be done in any language, several languages were specifically designed to produce structured programs. These languages force the user to use the main tools of program structuring, including code block structure using functions and subroutines, and data structure using records and sets. These languages include IBM's PL/1, Pascal, and C.

BASIC

Microsoft's BASIC interpreter comes installed in the IBM PC's *read-only memory* (ROM) and is also available on disk. The disk versions include the latest enhancements that have been made to BASIC. BASIC is easy to use and is one of the most powerful languages available for using all of the IBM PC's resources, including graphics, color, sound, communication through hardware ports (connection devices), and full-screen formatting. BASIC is provided at no extra cost with PC–DOS; no wonder it is the most popular programming language on the PC. Because the IBM PC ROM is proprietary, BASIC is always loaded from disk on IBM PC clones.

BASIC should be your first language if you are just learning to program. You can apply the many lessons learned while using BASIC to more advanced compiled languages, or you can purchase the BASIC compiler to make the programs written with the interpreter run faster. For real-time applications, the BASIC interpreter is not a good choice because it is difficult to add real-time drivers to interpreted BASIC programs. Figure 3.9a shows the integer-counting test program written in Microsoft QuickBASIC. This program will count to 32,767 using integers.

Pascal

Pascal was developed in 1971 by Niklaus Wirth, a Swedish computer science professor, who wanted to make learning to program easier. This programming language differed from other languages because he wanted to make it easy to write well-structured programs. Thus, Pascal was designed to force good structure on programmers using the language.

Pascal's popularity on the IBM PC and compatibles has very little to do with the teaching language originally developed. The prime reason Pascal is so popular on the PC is the introduction two years ago of Turbo Pascal by Borland International. Not only was Turbo Pascal attractively priced at $69.95, but it was loaded with features including full-screen formatting control, graphics, sound, and DOS calls. Turbo Pascal is a RAM resident program and requires no compilation time—a code can be entered and run immediately. Turbo Pascal was developed as a fast development tool for short- to medium-sized programs of 5000 lines of code or smaller. Larger programs must be cut into overlays. But this compromise is small compared with the huge advantages obtained with Turbo Pascal. Hundreds of thousands of people use the language, more than 300,000 at last count, which is more than all other Pascal compilers combined. Magazines publish articles about how to use it, and users groups and add-on products enhance application of the language.

The integer-counting program written in Turbo Pascal is shown in Figure 3.9b.

C Language

C is quickly becoming the language of choice among professional application and systems developers for the PC, other microbased systems, minicomputers, and mainframe computers. C was developed by Brian Kernighan and Dennis Ritchie as a programming language for the UNIX operating system. C is a block-structured language that contains some of the best features from other languages, including Pascal, PL/1, and Algol. UNIX itself was originally written in C for DEC PDP–11 minicomputers.

Although C was primarily designed for professional programmers, it can be used productively by beginners if approached cautiously. Doing simple things in C is easy, but mastering the language is difficult because a greater time investment is required on the part of the programmer than for easier languages like BASIC.

The integer-counting program written in Microsoft C is shown in Figure 3.9c.

FORTRAN

Historically, FORTRAN is the programming language of the scientist. Many scientific programmers still use the language, but most modern code writing on the IBM PC or compatibles is done in C. FORTRAN has not stood still. Over the years vast improvements have been made to the language that allow structured programming techniques to be practiced.

The integer-counting program written in Microsoft FORTRAN is shown in Figure 3.9d.

Assembly

No language is more arcane or difficult to learn and use than assembly language. But once the slow process of programming is complete, the results can be spectacular. The programs run incredibly fast, and all resources in the computer can be tapped with this language.

Individual instructions are represented by three- or four-letter mnemonics like MOV, JMP, MUL, and PUSH. When assembled, a program translates directly into machine code that the processor can read from memory and execute. Other higher level languages are also translated into machine code but usually in a less efficient manner that make these languages slower. Until a compiler can optimize a code better than a human, assembly language will always generate the tightest code and have the fastest execution. Only with assembly language can a person produce power programs such as Lotus 1–2–3 and the Turbo Pascal compiler and editor.

The integer-counting program written in IBM Assembler is shown in Figure 3.9e.

a

```
STIME$=TIME$
I%=0
WHILE I% < 32767
I%=I% + 1
WEND
ETIME$=TIME$
PRINT STIME$,ETIME$
STOP
```

b

```
PROGRAM INTCOUNT;
VAR X : INTEGER;
{$I SHOWTIME.SRC}
{$I ZEROTIME.SRC}
BEGIN
ZEROTIME;
X := 0;
WHILE X < MAXINT DO X:= X+1;
SHOWTIME;
END.
```

c

```
main ()
{
int i;
long t;
time(&t);
puts(ctime(&t));
i=0;
while (i < 32767)
i++;
time(&t);
puts(ctime(&t));
}
```

d

```
INTEGER COUNT
PRINT TIME
COUNT=0
DO 10 COUNT = 1,32767
COUNT = COUNT +1
10 CONTINUE
PRINT TIME
END
```

e

```
CSEG Segment Public 'CODE'
Assume CS:CSEG, DS:CSEG, ES:CSEG, SS:CSEG
Extrn StartTime:Near , PrintTime:Near
Org 0100h
Entry: Call StartTime ;External Subroutine
Sub AX,AX ;AX starts at 0
Test2Loop: Cmp AX,32767 ;Count until 32767
Je Test2End
Inc AX ;Otherwise increment
Jmp Test2Loop ;And do it again
Test2End: Call PrintTime ;External subroutine
Int 20h ;Exit
CSEG ENDS
END ENTRY
```

Figure 3.9 An integer-counting program in (a) BASIC, (b) Turbo Pascal, (c) Microsoft C, (d) Microsoft FORTRAN, and (e) IBM Macro Assembler code.

Products Mentioned in This Chapter

SideKick ($54.95 copy protected or $84.95 non-copy-protected), Borland International, 4585 Scotts Valley Dr., Scotts Valley, CA 95066. (408) 438–8400

PC/IX ($900), IBM, 220 Las Colinas Blvd., Irvington, TX 75062.

XENIX, Microsoft Corp., 10700 Northup Way, Box 97200, Bellevue, WA 98009. (206) 828–8080

Topview ($149), IBM, 1000 N.W. 51st St., Boca Raton, FL 33432. (305) 982–2690

Microsoft Windows ($99), Microsoft Corp., 10700 Northup Way, Box 97200, Bellevue, WA 98009. (206) 828–8080

GEM Desktop ($49.95), Digital Research Inc., 60 Garden Ct., Monterey, CA 93942. (408) 649–3896

Turbo Pascal ($69.95), Borland International, 4585 Scotts Valley Dr., Scotts Valley, CA 95066. (408) 438–8400

QuickBASIC Compiler ($99), Microsoft Corp., 10700 Northup Way, Box 97200, Bellevue, WA 98009. (206) 828–8080

BASIC Compiler ($495), IBM Entry Systems, 5201 South Congress Ave., Boca Raton, FL 33431. (305) 998–2000

Macro Assembler ($150), Microsoft Corp., 10700 Northup Way, Box 97200, Bellevue, WA 98009. (206) 828–8080

Microsoft C ($395), Microsoft Corp., 10700 Northup Way, Box 97200, Bellevue, WA 98009. (206) 828–8080

Microsoft FORTRAN ($350), Microsoft Corp., 10700 Northup Way, Box 97200, Bellevue, WA 98009. (206) 828–8080

The Norton Utilities ($99.95), Peter Norton, 2210 Wilshire Boulevard, Santa Monica, CA 90403. (213) 399–3948

Section TWO — Application Software

> *The ultimate dream of personal computing is that the PC will not only be incredibly versatile but become an extension of the mind.*
> —David Bunnell, Publisher of *PCWorld*

If laboratory robots extend the capabilities of our arms, hands, and fingers, then laboratory computers will become the extension of our minds. Even with current laboratory robots, a computerized control unit sends the desired sequence of signals to make each robotic movement. The commands and sequence of commands given to the robot arm, hand, and fingers is the software of the system.

This section will focus on application software of a more general type, but potentially just as useful for your work. With this software, you will be able to command robots of a different type; robots that print letters, plot graphs, and communicate information across the room or around the world.

Every type of application software is improving at a lightning pace. Application software is getting easier to use and more powerful in features and capabilities. I fully expect application software for the laboratory to provide systems that can extend our minds in the near future.

Chapter  FOUR

Word Processing

ord processing software was one of the first productivity software products written for PCs. Word processing software available for the IBM PC ranges from simple edit and print products to professionally engineered packages that outperform most stand-alone dedicated word processing systems.

Most word processors can create, store, edit, format, and print. *Creating* allows the user to type memos, letters, or reports with the characters that appear on the computer screen. The memo, letter, or report is now called a *document,* or *file. Storing* allows the user to store work on disk and to retrieve it in part or in whole at a later time. *Editing* allows the user to make changes to the text, such as fixing spelling errors and rearranging paragraphs or sentences. *Formatting* allows the user to see the document as it will appear when it is printed. Line justification, boldface print, subscripts, and superscripts can all be seen before printing begins. *Printing* allows the formatted document to be printed.

Word Processor Advantages

If you do any writing at all, from just a few letters and memos a week to huge reports, articles, or project proposals, then you can benefit from a word processor. If you have used a typewriter in the past, you will quickly appreciate the benefits of word processing. Word processors turn a computer into a super typewriter. The biggest advantage of a word processor is complete separation of the two main components in document production: the creative component where the content of the document is generated and the mechanical component where the document is physically printed (Figure 4.1). The mechanical component is also completely automated in a word processor, so the production of a document is fast and easy. The two components are connected in a typewriter, and if a misspelled word

1000–4/87/0073$06.00/1 © 1987 American Chemical Society

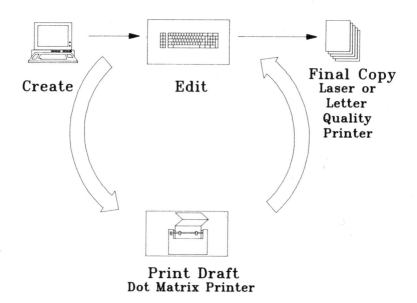

Create   **Edit**   **Final Copy**
Laser or
Letter
Quality
Printer

Print Draft
Dot Matrix Printer

Figure 4.1 Document generation using word processing software. The document is created, and a draft is printed on a dot-matrix printer. After the draft is edited on paper, the necessary changes are made on the screen with the word processor. The final version is printed with a letter-quality printer.

needs to be corrected, paragraphs reordered, or sentences rewritten, the entire page must be retyped. A word processor allows spelling mistakes to be fixed and any portion of the document to be edited before it is printed. Even after the first printing, the document can be edited and easily reprinted in part or whole. Many original copies of a document can also be made, so everyone who receives a letter will get a more personal, yet professional, impression. Using text-merging functions personalizes each document even more. *Text merging* allows a block of text to be substituted at a specific position in each printed document. The most common application of word processing is mass mailings, in which the same letter is sent to many people. The name, address, and salutation lines are the block of text that is substituted or merged for each letter, and each person gets an original copy of the letter.

Many scientists have found the quality of their writing to improve because of the ease and speed with which they can edit documents and correct grammar, punctuation, and spelling errors. Scientists can concentrate on the content of a document rather than on the mechanical work required to generate a document.

Because word processing programs are so popular, many types abound; you should be able to find one that fits your writing requirements and budget. Word processing programs are priced from $10 to more than $400. Most are easy to use and well-documented and perform as advertised. Some of the most popular word processors for the IBM PC are Microsoft Word, WordStar, Word Perfect, and Multimate. A popular, low-cost word processor is the product PC Write. Two word processors that were designed for scientific and technical writers are Spellbinder/Scientific and T3. Both of these programs allow chemical formulas, structures, and mathematical symbols to be used with normal text.

Another new class of text and graphics processors allows whole pages of output to be formatted and printed in camera-ready form on a laser printer. These systems are called desk-top publishing systems. With the improved near-typeset quality of text and graphics now available on laser printers, finished pages ready for reports, distribution, or slides can be generated with the appropriate software.

Word Processing Basics

To give a feeling for how word processing is performed, an interaction with a typical word processor will be presented. The program used in the following example is the "granddaddy" of word processors, WordStar. Program execution is begun by typing the program name at the PC–DOS prompt and pressing the Enter key.

$$C>WS$$

The initial screen shown in Figure 4.2 results. At this point the main function is selected from a menu of operations. If the user elects to edit a new document, the screen is transformed to that document, as shown in Figure 4.3.

On-Screen Editing

Now the text can be entered at any location on the screen. As characters are keyed in, the text is going into the computer's memory and is periodically written onto the disk. As the document gets larger, only a portion can be seen.

The document can be viewed approximately one page of text on the screen at a time by *scrolling* (Figure 4.4). The number of characters displayed on each line can be selected. When the end of a line is

```
                  editing no file
               < < N O - F I L E   M E N U > > >
 ---Preliminary Commands--- | ---File Commands--- | -System Commands-
 L Change logged disk drive |                     | R Run a program
 F File directory  on (OFF) | P Print a file      | X EXIT to system
 H Set help level           |                     |
 ---Commands to open a file-| E RENAME a file      | -WordStar Options-
 D Open a document file     | O COPY   a file      | M Run MailMerge
 N Open a non-document file | Y DELETE a file      | S Run SpellStar
```

Figure 4.2   A typical initial screen in a word processor. A function may be selected from the menu.

```
 A:RESUME.TXT  PAGE 1 LINE 12 COL 01          INSERT ON
 L----:----!----:----!----:----!----:----!----:----!----:----R

                 Glenn I. Ouchi, Ph.D.
 5989 Vista Loop    San Jose, CA  95124           (408) 723-0947
                                                  (408) 265-9243
 Education
     *  B.S. UCLA Chemistry
     *  M.S. University of Minnesota Organic/Biochemistry
     *  Ph.D. University of California, Santa Cruz
        Computer Applications in Chemistry
        Thesis Title: Computer-Assisted Prediction of
        Plausible Metabolites of Xenobiotic Compounds
```

Figure 4.3   The "what-you-see-is-what-you-get" screen on word processors allows line length to be set by selecting the end of a line on a "ruler". The page length and line spacing may also be selected.

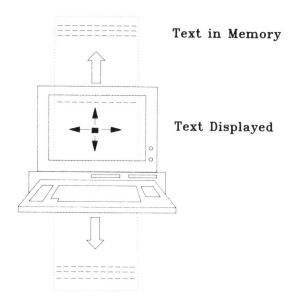

Text in Memory

Text Displayed

Figure 4.4 The display screen is a window that shows a portion of the text currently in memory.

reached, the text will automatically "wrap" around to the next line, and the text on the previous line will be spaced so the lines are justified. To *justify* is to make all lines the same width; thus, the left and right margins align. If you make an error, simply move the blinking line or cursor to the offending line, delete the error, and insert the correct text. Similarly, if you wish to insert some text, simply move to the desired location and start typing in the characters. If you need to rearrange some text, simply block off the text and either copy, move, delete, or write it onto the disk. To move around in the document, simply use the arrow keys or use the Page Up and Page Down keys to scroll backward and forward through the document. Word processors, such as Microsoft's Word, even display boldface type and underlines on the screen by graphic representations of all letters. To end the document creation session, the entire document is saved on disk.

Editing an Existing Document

A previously created document can be edited by selecting that function from the initial operation menu. The computer will then ask for the document file to be edited. The screen will display the document, and you will be able to use the cursor to enter, move, and delete text. If a section of this document is needed for another report, the section needed can be blocked off, written to disk, and read in the section from the disk into the new document.

Be certain that you are always editing a copy of the original document. Most word processors will make a back-up copy of a document automatically. If your word processor does not, be sure to make a copy before editing. Another problem can occur when the disk gets full. With WordStar, when the disk is full, the program does not allow the document that is being edited to be saved. One solution is to use "block" commands to identify the block of text just entered or edited and save just that block.

Printing the Document

The final step in document creation is printing. The printing function is selected from the initial operation menu. The document file to be printed is then selected, and the document is printed. To print a document correctly, the word processor must send the correct characters to the printer. When selecting a word processor, be sure the printer is supported by the word processing package.

Most printer features are supported by good word processing packages. These include page numbering, headers (lines at the very top), and footers (lines at the very bottom) on each page, boldface type, underlining, indentation, superscripts, subscripts, table formation, and margin control. Some of the new dot-matrix printers can print various fonts and character sizes with the appropriate software. New printers like the new low-cost laser printers have the ability to print different fonts and letter sizes. Word processing programs like Microsoft's Word and WordStar 2000 have the ability to control these additional printer capabilities. Printer control is performed by inserting special dot commands in WordStar. These commands are called *dot commands* because they are preceded by a period, or dot.

Word or Character Combination Search and Replace

Most word processing packages have the ability to search the entire document for a specific word or set of characters (Figure 4.5). This search can be extremely useful when looking for or replacing specific words, for example, replacing every occurrence of *p*-bromochloro-benzene with *o*-dichlorobenzene. Programmers can make use of this

```
^Qa    A:RESUME.TXT    PAGE 1 LINE 10 COL 64              REPLACE (Y/N):

^S=delete character   ^Y=delete entry    ^F=File directory
^D=restore character  ^R=Restore entry   ^U=cancel command

    FIND? o    REPLACE WITH? %&*    OPTIONS? (? FOR INFO) g

L--!--!--!--!--!--!--!--!--!--!--!--!--!-----------------R
.CW 10
.LH 8
.HE Auth%&*r Inf%&*rmati%&*n                        Page #
.FO Glenn I. Ouchi                                  Resume

            ^BGlenn I. Ouchi, Ph.D.^B
5989 Vista L%&*%&*p    San J%&*se, CA  95124
                                        (408) 723-0947
                                        (408) 265-9243

^BEducati%&*n
^B   *   B.S. UCLA Chemistry
     *   M.S. University %&*f Minnes%&*ta Organic/Bi%&*chemistry
     *   Ph.D. University %&*f Calif%&*rnia, Santa Cruz
         C%&*mputer Applicati%&*ns in Chemistry
         Thesis Title: C%&*mputer-Assisted Predicti%&*n %&*f
         Plausible Metab%&*lites %&*f Xen%&*bi%&*tic C%&*mpounds
```

Figure 4.5   An example of search and replace. The screen shows the result of replacing all occurrences of the letter o with %&.*

feature while debugging codes because the search will find every location in the program where different variable names are used.

Special features of some advanced word processors include the ability to undo one or more steps performed, to format the printed results in two or more columns, and to provide automatic headers, footers, indexes, and tables of contents.

Spelling Checkers and Dictionaries

Most word processing packages have the ability to use spelling checkers (Figure 4.6). Spelling checkers compare each word in a document against a dictionary of 10,000–30,000 words. Words not found in the dictionary are flagged and presented to the user to either respell, delete, or add to a user's dictionary. Spelling checkers ensure that words are valid combinations of letters but cannot catch misspellings when the misspelled word is also a valid word. For

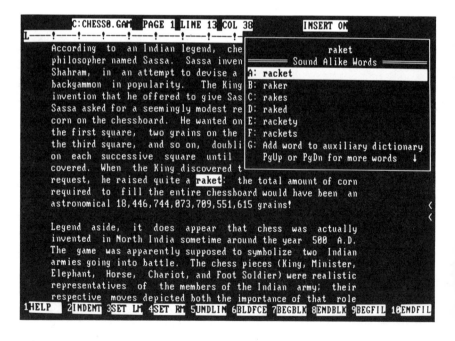

Figure 4.6 Detected misspelled words are displayed with various options to correct the spelling, add the word to a user dictionary, or ignore the word. (Photo courtesy of Borland International.)

example, if the number "four" is misspelled "for", the misspelling will not be detected. Some advanced spelling checkers even use phonetic spelling of words and provide the correct spelling for the user to select.

Distribution-List Merging and Boiler Plate Documents

Most advanced word processors have a capability to merge distribution lists or mailing lists into form documents or letters (Figure 4.7). As each document is printed, the appropriate name, location or address, and other personalized information is printed into the document.

Boiler plate documents are created by merging whole paragraphs or sections of text to create a custom-made document (Figure 4.8). In other words, once you have typed in some text and you like your work, you need not retype the text. These techniques save time and typing errors.

Scientific Word Processors

If a standard word processing program cannot fill your scientific or technical needs, then you should evaluate one of the many special scientific versions. These programs include the standard word processing capabilities along with the ability to display and print mathematical equations, Greek symbols, and chemical formulas. With the improved graphics screens, printers, and computational speed of the PC/AT and clones, these products are now very attractive. A list of some current word processors that fall in this category is provided at the end of this chapter.

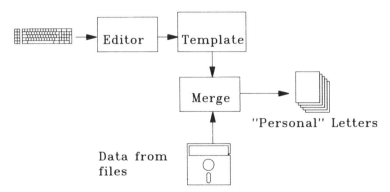

Figure 4.7   Steps involved in preparing a set of documents by using a merge list of information.

PC MEMORY

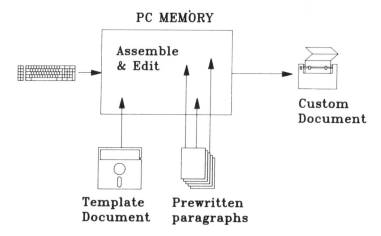

Figure 4.8   Steps involved in preparing a document by using a prewritten boiler plate section of text and a template document.

Desk-Top Publishing Systems

Most chemists need more than just text processing to create finished reports and documents. Two recent additions to IBM PC and compatibles have made the creation of finished document pages with both text and graphics a reality within financial reach. First, low-cost laser printers with the ability to print near-typeset-quality text with graphics has made the creation of the final pages available at a reasonable cost. Second, software that formats combinations of text and graphics continues to improve. For example, Aldus Page Maker is page-composition software capable of formatting and printing a full page of text on the Apple Laserwriter printer. Page formatting includes multiple columns of text; font size and style switching; and boxes, lines, and flowing text around figures and symbols. This software also includes the ability to include graphs and to plot the graphs at a resolution of 300 dots per inch. An example of a complete formatted page is shown in Figure 4.9. Currently, this type of software is available only on the Apple Macintosh using an Apple Laserwriter printer. However, improved software is promised in late 1986, including a version of Page Maker from Aldus for the IBM PC.

Laboratory PC User

Newsletter for the Laboratory PC Users Group

Lotus in the Lab

Maintaining the Control Chart Data

Lotus 1-2-3 data management commands can be used to maintain your control chart data.

Each time a control sample is analyzed the results of the analysis can be entered into your Lotus data base. A portation of a typical data base for a chromatographic analysis is shown in Figure 2. This data b·se can easily be modified to track data from any type of instrument in your laboratory. If your instrument can communicate with a PC, or has a PC based data system, this data could be captured and placed into Lotus without having to re-key in the data. Techniques for capturing data will be described in a future article in this column.

What is a Data Base?

A data base is a structured collection of information. A telephone book is a common example of a printed data base with thousands of entries. Each entry in the telephone book usually has three pieces of information: a name, an address and a telephone number. In data base management terminology, the entries are called records. The pieces of information contained in each record are called fields. Thus, a telephone book has thousands of records, each with information in one of three fields: a name, an address and a telephone number.

We are all familiar with printed data bases and use them successfully in our daily lives. Most are easy to use. Here are some examples of other printed data bases we use often with their normal fields in parenthesis: the card catalog at the library (book title, author, subject), restaurant menu (food item, price), store merchandise catalog (item, description, price), and Merck Index (compound name, formula, description, other physical, pharmacological and chemical data).

Computerized data bases are no more difficult to use than are printed data bases. Computerized data bases have advantages over printed data bases since the access to the data is not limited to a static method of sorting and presenting the data. For example, a telephone book presents its data sorted by the names in alphabetical order. If you know a person's name, you can find their telephone number (as long as it is listed and you are looking in the right telephone book !). But a telephone book is pretty

useless if you only have a telephone number and need to know who's number it is. A computerized data base could easily solve this problem by sorting the telephone numbers in numerical order or by making a query to the data base for a direct match. Creating a computerized data base thus provides you with the ability to access your data using many different criteria and sorting capabilities.

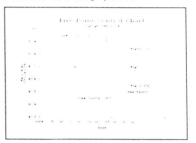

Figure 1. Data Control Chart generated using Lotus 1-2-3.

Creating a Control Chart Data Base

Our control chart data base contains the data from the analysis of a number of different control samples over a two week period. The results are from three different instruments, gas chromatograph one (GC1), gas chromatograph two (GC2) and liquid chromatograph one (LC1) for two different compounds, butane and mparabin. In an actuai application you may want to create separate data bases for each instrument or even each compound

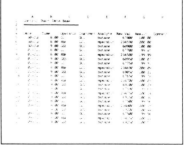

Figure 2. Data Base for control chart data.

Figure 4.9   Complete printed page from Aldus Page Maker software. This software can mix text and graphics and print them on the Apple Laserwriter laser printer.

Products Mentioned in This Chapter

Microsoft Word ($375), Microsoft Corporation, 10700 Northup Way, Box 97200, Bellevue, WA 98009. (800) 426–9400

WordStar and WordStar 2000 ($495), MicroPro International Corp., 33 San Pablo Ave., San Rafael, CA 94903. (415) 499–1200

Multimate ($495), Multimate International Corp., 52 Oakland Ave. N, East Hartford, CT 06108. (203) 522–2116

PC Write ($10), Quicksoft, 219 N. First #224, Seattle, WA 98109. (206) 282–0452

Word Perfect ($495), Satellite Software International, 288 W. Canter St., Orem, UT 84057. (801) 224–8554

Scientific Word Processors

T3 ($595), TCI Software Research Inc., 1190B Foster Road, Las Cruces, NM 88001. (800) 874–2383

Spellbinder/Scientific ($695), Lexisoft, Inc., P.O. Box 1378, Davis, CA 95617. (916) 758–3630

Brit Scientex ($795), Scientific Communication Corp., 2136 Locust St., Philadelphia, PA 19103. (215) 732–7978

Proofwriter ($425), Image Processing Systems, 6409 Appalachian Way, Madison, WI 53705. (608) 233–5033

Volkswriter Scientific ($495), Lifetree Software Inc., 411 Pacific St., Monterey, CA 93940. (408) 373–4718

Desk-Top Publishing

Aldus Page Maker, Aldus Corporation, 411 First Ave. South, Seattle, WA 98104. (206) 622–5500

Apple Macintosh and Apple Laserwriter printer, Apple Computer, Inc., 20525 Mariani Ave., Cupertino, CA 95014. (408) 996-1010

Chapter  Five

Spreadsheets

f one program had to be selected that sparked the beginning of the PC revolution, that program would be VisiCalc, the spreadsheet program written by Software Arts and initially marketed by VisiCorp. (To show how quickly the fortunes of a software company can change, Software Arts and VisiCorp no longer exist. Software Arts was purchased by Lotus Development Corporation in 1985, and VisiCorp merged with Paladin Software Corporation in 1984. The rights to VisiCorp programs were recently sold to Control Data Corporation.) VisiCalc, more than any other program, brought the PC out of the hobbyist's workshop and into the modern corporate office. For the first time, a user with no programming skills could take full advantage of the resources of a computer. VisiCalc gave the nonprogrammer the ability to create a financial model and easily manipulate it. The model could be stored, recalled, and updated in a few minutes. New entries could be made into the model, and the whole model could be recalculated in a few seconds. With the press of a few keys, the model or any parts of it could be printed or even made into a graph using VisiTrend or VisiPlot.

Before VisiCalc, this sort of personal computing tool was nonexistent; calculations and even text and files for these models had to be retyped and recomputed. This process could easily take hours or days. After VisiCalc, PCs began to appear in offices across the country and around the world.

The most popular program for the IBM PC is Lotus 1–2–3, an integrated application program that contains one of the most advanced spreadsheets as its central application. A *spreadsheet* program displays a grid in which data can be input. Various calculations can then be performed by the program. Using the spreadsheet, a user can enter or import data, generate reports and graphs, and perform data management tasks. Spreadsheets continue to have a dominating influence over the use of PCs. Because of new

1000–4/87/0085$06.50/1   © 1987 American Chemical Society

versions such as Lotus 1–2–3 version 2 and newly developed models such as VP-planner and Javelin, spreadsheets have a bright future. Integrated applications are also available with additional functions such as word processing and data communications; examples include Framework II and Lotus Symphony.

Spreadsheet Applications for the Laboratory

Spreadsheet applications do not stop at the accounting office doors. The features can just as easily find applications in research, development, and quality control laboratories. The basic function of these programs is to provide an electronic *worksheet* that allows the user to quickly and easily set up a mathematical model, enter data and text into the model, and report the results. Any task currently done with paper, pencil, and a calculator can be done more easily by using a spreadsheet program. A spreadsheet program allows the computer screen to become your paper, the computer keyboard to become your pencil, and the computer to become your calculator. An even more convenient application is that an entire worksheet can be stored to work on later or to archive the information. Stored data can also be consolidated into a summary report. Once the data is in computer-readable form, many ways to analyze and report the data are possible.

Three major elements of all spreadsheets make these programs popular and useful as research tools. First, the screen setup of rows and columns resembles the grid found on the pages in most research notebooks. Second, the philosophy of "what-you-see-is-what-you-get" (or WYSIWYG) allows direct construction of the desired report. There are no abstract requirements to enter data or labels; simply move to the desired location and type in the entry. What is seen on the screen is the report that will be printed on the printer or disk file. Finally, the program gives instantaneous feedback after every entry. If the location of a column of values in a report isn't satisfactory, move the column; if an erroneous entry is made, the entry will be seen immediately and can be corrected.

The electronic worksheet is organized as a grid of columns and rows (Figure 5.1). Each column is identified by a capital letter, and each row is labeled with a number. At each intersection of a row and column are entry positions, or *cells*, that can contain information. The cells are identified by the column letter and row number that intersect the cell's location on the screen.

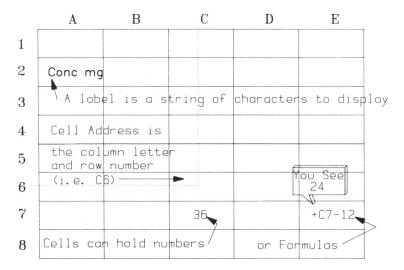

Figure 5.1 Spreadsheet input screen shows where text or numbers can be entered on the screen in grid locations where columns and rows meet. Each grid location is called a cell.

Five types of information can be placed in a cell:

1. numbers (e.g., 123, –12.34, or 0.09876);
2. text or titles (e.g., ACME Research & Development Labs);
3. formulas (e.g., **100*C3**, which means to multiply the contents of cell C3 by 100);
4. preprogrammed functions (e.g., **@AVG(range of cells)**, which means to display the average of the given range of cells); and
5. commands for macros.

Entering information into the worksheet is done by moving a lighted box (*cursor*) to the desired grid location and typing in the entry. On numerical data, the basic math functions (addition, subtraction, multiplication, and division) can be performed, and many other functions can be preprogrammed for easy execution. One strength of spreadsheet programs is that when a value is changed in the worksheet, all values that are computed from that particular value are instantly recomputed and the results displayed. This capability allows "what if" calculations to be performed in just a few moments.

The spreadsheet also features the ability to copy or move the contents of one cell to other cells. Moving cells allows data to be displayed exactly in the desired way. If a column of values has been entered in the wrong place on the worksheet, the values can easily be moved.

These simple rules allow complex worksheets to be developed and used. Because the entire worksheet can be stored, templates of calculations and reports can be built. Once a report template has been prepared, anyone can simply key in new raw data values and generate a new report from the current data.

The mechanics of data entry and reporting can also be automated by using macros. *Macros* allow several keystrokes to be assigned to a single keystroke. Macros also have programmable steps, so logic can be built into the macro execution. Macros are powerful tools because they can automate a predominantly manual application.

The utility of spreadsheet programs can be illustrated best by some examples. The examples in this chapter were prepared in one of the first integrated programs, Lotus 1-2-3. Lotus 1-2-3 combines a spreadsheet with data management and graphics. At the heart of Lotus 1-2-3 is the spreadsheet. All data is entered in the spreadsheet, whether the data is for the spreadsheet, data management, or graphics.

Computing Chromatographic Results

Spreadsheets can be used for the analysis of chromatographic data. Three types of chromatographic analysis reports will be generated in this chapter: area percent, single-level standard quantitation using external standards, and multilevel standard quantitation using a linear least-squares fit to the standard data. The advantage of generating these reports using a spreadsheet is that any format can be created and any needed calculations can be performed by adding to the existing report.

Area Percent

An *area percent report* computes the percentage of total sample that each component in a chromatographic run represents. The report assumes the detector response is identical for each component. The area percent report is generated by summing the areas or heights of each peak to compute the total area of all peaks. Then the area of each individual peak is divided by the total area and multiplied by 100. Figure 5.2 shows a typical area percent report as a Lotus 1-2-3 application.

To generate this area percent report by using Lotus 1-2-3, the following instructions must be followed.

```
       A          B          C          D          E          F          G
1                  Liquid Chromatograpy Data
2                    Area Percent Report
3         ═══════════════════════════════════════════════
4      Sample:   QA #2345            Date:      04-Mar-85
5      Column:   Sephadex            Operator:GIO
6      Solvent:  Methanol/Water      Inst No.:LC #1234
7      Notes:    Sample vial seal broken during shipment
8         ═══════════════════════════════════════════════
9      Peak No.  Ret. Time   Area    % Total Area
10        1         2.34     12345      1.160
11        2         2.67     34567      3.248
12        3         3.68     78904      7.414
13        4         4.98      5674      0.533
14        5         6.89     12356      1.161
15        6         9.78    896654     84.249
16        7        10.76      9223      0.867
17        8        13.45     14568      1.369
18                          ------------------
19                          1064291   100.000
20
```

Figure 5.2   Typical area percent report generated in Lotus 1–2–3 that reports chromatographic data.

Follow the step-by-step instructions for loading Lotus 1–2–3. First, make sure the worksheet has nothing in it by erasing the entire worksheet with the command / **Worksheet Erase Yes**.

A blank worksheet containing only a column of numbers and a row of letters should be seen. Now the section for the report header information can be created.

- Move the cursor to cell B1 and type **Liquid Chromatographic Data**.
- Move to cell C1 and type **Area Percent Report**.
- Move to cell A3 and type \=. This will fill the cell with equal signs. This pattern can be made across the entire row by copying the contents of this cell to the other cells in the row with the command / **Copy A3<ENTER>B3..E3<ENTER>**.
- Now enter the header labels in these cells in columns A and D. In A4 enter **Sample:**, in A5 enter **Column:**, in A6 enter **Solvent:**, in A7 enter **Notes:**, in D4 enter **Date:**, in D5 enter **Operator:**, in D6 enter **Inst No.:**.
- Enter another equal sign pattern in row 8 by entering \=. Copy this to the other cells with / **Copy A8<ENTER>B8..E8<ENTER>**.

This procedure generates a report header. Making a few changes to this header allows it to be used on many different reports. Saving the work enables this header to be used again and also ensures that the work done thus far won't be lost. To save the work, enter /**File Save HEADER**<**ENTER**>.

- Now enter the column headings for the report in row 9. In A9 enter **Peak No.**, in B9 enter **Ret. Time**, in C9 enter **Area**, and in D9 enter **% Total Area**.
- Enter the peak numbers, retention times, and areas as shown:

	A	B	C
10	1	2.34	12345
11	2	2.67	34567
12	3	3.68	78904
13	4	4.98	5674
14	5	6.89	12356
15	6	9.78	896654
16	7	10.76	9223
17	8	13.45	14568

- Place the underlines in cells C18 and D18 with \-. Now some calculations can be done. The percent total area of each peak can be computed, but the total area of all of the peaks in the run is needed. This total area can be computed by using the **@SUM** function. In cell C19, enter +**@SUM(C10..C17)**. Note that the range of cells to sum can be interactively selected by using the Lotus pointing features.
- Now move to cell D10 and enter the equation +**100*C10/C19**. This equation will compute the percent area of the first peak. This equation can be copied to the other cells in column D with /**Copy D10**<**ENTER**>**D11..D17**<**ENTER**>. Notice that cells D11–D17 have **ERR** displayed. This is the Lotus 1–2–3 error message, which means an erroneous equation has been entered. Move to cell D11 and view the contents, which should be **100*C11/C20**. The error is generated because cell C20 has no value, and Lotus is dividing by zero. But why did Lotus 1–2–3 even consider using cell C20 when cell C19 was the desired cell? The manner that Lotus 1–2–3 copies the contents of cells answers this question. Lotus 1–2–3 helps the user whenever it can. Usually, if the contents of a cell or range of cells is copied to other cells, the cell letters or numbers used in equations must be incremented. In the previous example, the C column cell number was correctly incremented from the original C10 to C11 in the new equation. The problem, division by zero, occurred because the other cell identifier, C19, was also incremented to C20.

This problem can be solved by using *absolute references* in a cell identifier. Placing a $ in front of a row or column identifier, or both the row and column identifiers, marks it as an absolute reference and indicates to Lotus 1–2–3 not to increment the cell reference when copied or moved.

Thus, the correct equation for cell D10 is +**100*C10/\$C\$19**. When this equation is copied to the other cells in column D, the cell identifiers for the areas will be incremented by the total area cell, and C19 will not be incremented. Now the cell can be copied with the command /**Copy D10<ENTER>D11..D17<ENTER>**.

To finish this report, sum the **% Total Area** column within cell D19 by entering the function +**@SUM(D10..D17)**.

Single-Level Standard Quantitation

Rather than assuming the detector response is the same for each compound to be quantitated in a chromatographic run, known concentrations of standards can be analyzed and the actual response used to calculate the concentrations of specific components in samples. The standards are run under the same conditions, and the areas or heights are recorded. Then for each compound, a detector response factor is computed. The *response factor* is the compound's detected area or height divided by its concentration in the standard. To compute the concentration of a compound in an unknown sample, simply divide the area of the compound in the sample by its response factor (Figure 5.3).

Reports contain standard calibration data and computed sample results. An example is shown in Figure 5.4. This report can be generated if the following procedure is followed.

The report generation is started by making some small modifications to the header of information created in the area percent worksheet. This worksheet is retrieved by using the command /**File Retrieve HEADER<ENTER>**.

The display will look like that shown in Figure 5.5. Now the second line in the report header can be changed. First erase the contents of cell C1. Place the cursor in cell C1 and enter /**Range Erase C1..C1<ENTER>**. Now move the cursor to cell B1 and enter the label **Single-Level External Standard**.

Enter the following new column headings in row 9: in A9 enter **Compound**, in B9 enter **Ret. Time**, in C9 enter **Area**, in D9 enter **Amount**, in E9 enter **Resp Factor**, in F9 enter **Amt %**, and in G9 enter **RT Delta%**.

	A	B	C	D	E	F	G
1		Liquid Chromatograpy Data					
2		Single Level External Standard					
3	=====	=====	=====	=====	=====	=====	=====
4	Sample:	QA #2345		Date:	04-Mar-85		
5	Column:	Sephadex		Operator:GIO			
6	Solvent:	Methanol/Water		Inst No.:LC #1234			
7	Notes:	Sample vial seal broken during shipment					
8	=====	=====	=====	=====	=====	=====	=====
9	Compound	Ret. Time	Area	Amount	Resp Factor	Amt %	RT Delta%
10	Methane	2.34	12345	52.630	234.56	1.317%	0.426%
11	Ethane	2.67	34567	99.936	345.89	2.501%	0.000%
12	Propane	3.68	78904	336.076	234.78	8.411%	0.271%
13	Butane	4.98	5674	23.954	236.87	0.600%	0.000%
14	Pentane	6.89	12356	50.303	245.63	1.259%	0.000%
15	Hexane	9.78	896654	3334.030	268.94	83.444%	-0.205%
16	Heptane	10.76	9223	39.319	234.57	0.984%	0.186%
17	Octane	13.45	14568	59.299	245.67	1.484%	0.148%
18			-----------------			---------	
19			1064291	3995.548		100.00%	
20							

Figure 5.3   The concentration of a compound can be computed from its response factor. The response factor is computed from a known standard concentration of the compound.

Before the sample report is completed, the calibration data section must be set up. Move the cursor to A22 and enter **Calibration Data**. Enter these labels for the column headings in row 23: in A23 enter **Compound**, in B23 enter **Ret. Time**, in C23 enter **Area**, in D23 enter **Amount**, and in E23 enter **Resp Factor**.

The calibration data can now be entered as follows:

	A	B	C	D
	A	B	C	D
24	**Methane**	2.34	23456	100
25	**Ethane**	2.67	34589	100
26	**Propane**	3.69	23478	100
27	**Butane**	4.98	23687	100
28	**Pentane**	6.89	24563	100
29	**Hexane**	9.76	26894	100
30	**Heptane**	10.78	23457	100
31	**Octane**	13.47	24567	100

The response factor is computed in column E by entering the formula +C24/D24, which divides the standard area by the standard amount. Copy this formula to the other cells in column E with /Copy E24..E24<ENTER> TO E25..E31<ENTER>. Now the sample data report can be completed. Enter this data in columns A, B, and C.

```
Liquid Chromatograpy Data
Single Level External Standard
```

Sample:	QA #2345		Date:	04-Mar-85
Column:	Sephadex		Operator:GIO	
Solvent:	Methanol/Water		Inst No.:LC #1234	
Notes:	Sample vial seal broken during shipment			

Compound	Ret. Time	Area	Amount	Resp Factor	Amt %	RT Delta%
Methane	2.34	12345	52.630	234.56	1.317%	0.426%
Ethane	2.67	34567	99.936	345.89	2.501%	0.000%
Propane	3.68	78904	336.076	234.78	8.411%	0.271%
Butane	4.98	5674	23.954	236.87	0.600%	0.000%
Pentane	6.89	12356	50.303	245.63	1.259%	0.000%
Hexane	9.78	896654	3334.030	268.94	83.444%	-0.205%
Heptane	10.76	9223	39.319	234.57	0.984%	0.186%
Octane	13.45	14568	59.299	245.67	1.484%	0.148%
		-------	-------		---------	
		1064291	3995.548		100.00%	

```
Calibration Data
```

Compound	Ret. Time	Area	Amount	Resp Factor
Methane	2.35	23456	100	234.56
Ethane	2.67	34589	100	345.89
Propane	3.69	23478	100	234.78
Butane	4.98	23687	100	236.87
Pentane	6.89	24563	100	245.63
Hexane	9.76	26894	100	268.94
Heptane	10.78	23457	100	234.57
Octane	13.47	24567	100	245.67

Figure 5.4   A sample chromatographic report contains the standard calibration data and computed sample results.

	A	B	C
10	**Methane**	**2.34**	**12345**
11	**Ethane**	**2.67**	**34567**
12	**Propane**	**3.68**	**78904**
13	**Butane**	**4.98**	**5674**
14	**Pentane**	**6.89**	**12356**
15	**Hexane**	**9.78**	**896654**
16	**Heptane**	**10.76**	**9223**
17	**Octane**	**13.45**	**14568**

Place underlines in cells C18, D18, and F18. Then sum the C column in cell C19 with the entry +**SUM(C10..C17)**. Similarly, sum the D column in cell D19 with +**SUM(D10..D17)**, and the F column in cell F19 with +**SUM(F10..F17)**.

B2:   Area Percent Report

```
       A        B        C        D        E          F        G
 1              Liquid Chromatograpy Data
 2                 Area Percent Report
 3         ====================================================
 4   Sample:  QA #2345          Date:      04-Mar-85
 5   Column:  Sephadex          Operator:GIO
 6   Solvent: Methanol/Water    Inst No.:LC #1234
 7   Notes:   Sample vial seal broken during shipment
 8         ====================================================
 9
10
11
12
13
14
15
16
17
18
19
20
```

Figure 5.5   Spreadsheet with header information.

The other columns are generated by formulas. In cell D10, enter +**C10/E10**, which is the peak area divided by the response factor. In cell E10 enter +**E24**, the contents of the response factor computed from the calibration standard data. Cell F10 computes the percentage of this compound's amount of the total amounts of all compounds in the sample. This formula is +**D10/$D$19**. Note the use of the absolute reference by the **$** in front of D and 19. The contents of cell D19 will be used exclusively, even when this formula is copied or moved. Finally, G10 is the percent difference in retention time and is computed with the formula +**(B24-B10)/B24**.

The formulas in row 10 can now be copied to the other rows with the copy command /**Copy D10..G10**<ENTER> **TO D11..G17** <**ENTER**>. The single-level external standard report should now appear as shown in Figure 5.4.

Multilevel Standard Quantitation

Applying the coefficients of a linear least-squares fit to a set of calibration data is common in many laboratory applications. Normally, standards are prepared at various concentrations and analyzed. The standard results are pairs of values, concentration and response (area or height in chromatographic data, absorbance in UV spectroscopy) that can be fit to a line, or another mathematical function, making a calibration curve. To compute the concentration of a component in a sample, the detected response is entered into the calibration curve equation.

Lotus 1–2–3 can compute the calibration curve and also plot the calibration points and least-squares fit line. (Recently, Lotus 1–2–3 version 2 added a new command called *data regression*. This command makes performing regression analysis in Lotus 1–2–3 version 2 much simpler than in earlier versions.)

The equations for computing a linear least-squares fit are

$$Y = a + bX$$

$$b = \frac{\sum X_i Y_i - \dfrac{\sum X_i \sum Y_i}{n}}{\sum X_i^2 - \dfrac{(\sum X_i)^2}{n}}$$

$$a = \left(\frac{\sum Y_i}{n} - b \frac{\sum X_i}{n} \right)$$

These calculations can be entered into the spreadsheet and used as is, but by using a little spreadsheet strategy, a very flexible model can be created.

First, the location of the X and Y values on the spreadsheet must be planned and referred to. Lotus 1–2–3 provides the ability to name a range of cells. The range of cells can then be used in equations by a descriptive name rather than only as a range of cells. The concentration values in column A and the area values in column B must be placed. The calibration standards will be at five levels from 100 to 500 with their corresponding area values. If these values and labels are entered, the spreadsheet will look like Figure 5.6.

Part of the least-squares fit calculation uses the product of the concentration and the area at each level. A separate column of these calculations can be made in column C. In cell C6, the equation +A6*B6 can be entered. Now this equation will be copied to the rest of the column with the command /Copy C6..C6<ENTER> TO C7..C10<ENTER>.

	A	B	C	D
1	Calibration Data for 1,3 dichlorobenzene			
2				
3	Concentration	Area		
4				
5				
6	100	21000		
7	200	30400		
8	300	39700		
9	400	48900		
10	500	61000		

Figure 5.6   First step in multilevel standard quantitation using
a linear least-squares fit to the standard data.

The program automatically increments the cell references to match
the copied rows. Copying the cells in this manner is necessary for this
application.

Now the cell ranges used in the calculations can be named. First
the concentration range of cells will be named CONC with the
command /**Range Name Create CONC**<ENTER> **A6..A10**
<ENTER>. Then the area and product of area and concentration
ranges will be named AREA and PRODUCT, respectively, with /**Range
Name Create AREA**<ENTER> **B6..B10**<ENTER> and /**Range Name
Create PRODUCT**<ENTER> **C6..C10**<ENTER>. Now the slope and
Y-intercept equations for the least-squares fit line can be entered. The
entries and formulas are as follows: in A13 enter **Slope of line:**, in B13
enter +(**@SUM(PRODUCT)** / **@COUNT(CONC)@AVG(CONC)***
@AVG(AREA)) / **@STD(CONC)^2**, in A14 enter **Y-Axis Intercept:**,
and in B14 enter +**@AVG(AREA)B13*@AVG(CONC)**. The resulting
worksheet will look like Figure 5.7.

In Chapter 6, this data and the least-squares fit line will be plotted.
Now some "what if" calculations will be performed with the model.
By changing any of the concentration or area values, the slope of an
intercept will change. Thus, if the area for the 100 concentration
standard is changed to **14000** and if the change is made in cell B6,
then all values will be recalculated, and a new slope and intercept will
be displayed as shown in Figure 5.8.

Expanding the Number of Standard Levels.   Setting up the
spreadsheet as shown by using named ranges rather than fixed ranges
allows the number of levels to be changed easily. This change can be
made by adding more data in the A and B columns, adding equations

	A	B	C	D
1	Calibration Data for 1,3 dichlorobenzene			
2				
3	Concentration	Area		
4				
5				
6	100	21000		
7	200	30400		
8	300	39700		
9	400	48900		
10	500	61000		
11				
12				
13	Slope of Line:	98.5		
14	Y-Axis Intercep	10650		
15				
16				
17				
18				
19				
20				

Figure 5.7   Least-squares fit spreadsheet.

	A	B	C	D
1	Calibration Data for 1,3 dichlorobenzene			
2				
3	Concentration	Area		
4				
5				
6	100	14000	1400000	
7	200	30400	6080000	
8	300	39700	11910000	
9	400	48900	19560000	
10	500	61000	30500000	
11				
12				
13	Slope of Line:	112.5		
14	Y-Axis Intercep	5050		
15				
16				
17				
18				
19				
20				

Figure 5.8   "What if" calculation performed by changing the area value of the 100 standard from **21000** to **14000**.

to the C column, and changing the named ranges. To add two new values, first insert two more rows of cells by starting in row 11. Use the command /**Worksheet Insert Row {Down Arrow}<ENTER>**. Then these additions can be made to the worksheet: in A11 enter **600**, in A12 enter **700**, in B11 enter **697000**, and in B12 enter **801000**.

The product in column C can be copied with the command /**Copy C10..C10<ENTER> TO C11..C12<ENTER>**. The range names can now be changed to include the new values; the commands are /**Range Name Create CONC<ENTER> A6..A12<ENTER>**, /**Range Name Create AREA<ENTER> B6..B12<ENTER>**, and /**Range Name Create PRODUCT<ENTER> C6..C12<ENTER>**. Now the slope and intercept values will calculate the correct values as shown in Figure 5.9.

Making a Template. If the model is used as a template, then a completely new calibration curve can be generated by entering new concentration and area values. A linear least-squares fit application

	A	B	C	D
1	Calibration Data for 1,3 dichlorobenzene			
2				
3	Concentration	Area		
4				
5				
6	100	14000	1400000	
7	200	30400	6080000	
8	300	39700	11910000	
9	400	48900	19560000	
10	500	61000	30500000	
11	600	69700	41820000	
12	700	80100	56070000	
13	Slope of Line:	106.5		
14	Y-Axis Intercep	6514.2857143		
15				
16				
17				
18				
19				
20				

Figure 5.9 Result of two new values added to the original least-squares fit.

that can be used with any data has thus been generated. Applications like this one can help solve many problems. Many types of templates can be created and used.

Using Preprogrammed Functions: Quality Data Charts. To ensure the quality of the data and in turn the validity of the information that analyses provide, the scientist must document how an analysis is performed and provide historical results of the analysis on similar samples. This information is usually provided in the form of a quality data or quality control chart such as the chart shown in Figure 5.10.

This chart was created from built-in calculation functions such as **AVG**, which is average, and **STD**, which is standard deviation. The original set of results can be entered into column A. Then the average and one standard deviation can be computed. This data can also be plotted and will be shown in the graphics chapter.

Enter these values in column A: in A5, **23456**; A6, **23789**; A7, **23333**; A8, **24090**; A9, **23567**; A10, **23890**; A11, **23897**; A12, **24444**; A13, **23555**; A14, **23666**; and A15, **23678**.

Now the average and standard deviation can be computed. In cell A20 enter +**@AVG(A5..A15)**, and in A21 enter +**@STD(A5..A15)**.

These examples are just the tip of the iceberg for applications. Many more useful applications will be presented throughout this book.

Data Importing:
Getting Data into a Worksheet from Disk Files

Data used in the spreadsheet is not limited to data entered through the keyboard. In a number of ways, data can be moved into a Lotus worksheet if data is already stored in PC–DOS disk files. The easiest way is to give the file name the extension **.PRN**, then import the data using the /**File Import** command with either **Numbers** or **Text**. The Numbers option will bring in text numbers and place them in the worksheet as normal numbers. This is the most popular way to enter data from a foreign file. The Numbers option ignores all labels and does not read in any text. If only the text from a file is desired, then the Text option can be used.

The data captured from an instrument and stored in a disk file can be easily imported into Lotus. The extension on the disk file must be set to **.PRN**. Only files with the **.PRN** extension can be imported into

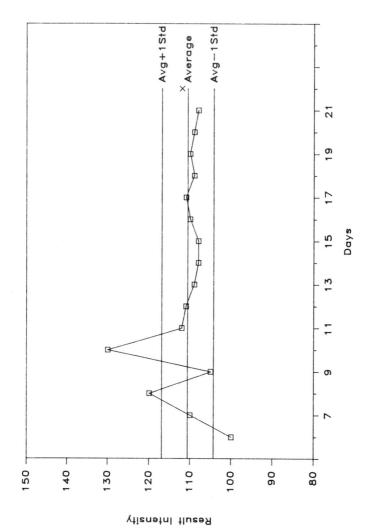

Figure 5.10   Instrument quality control chart. The graph shows the results of check samples that have the same concentration analyzed once per day. The lines connect the previous sample values. The single point is the current data point.

Lotus 1–2–3 version 1a, with the command /**File Import Numbers FILENAME**<**ENTER**>. Lotus 1–2–3 version 2 files with any extension can be imported with this command. This command enters the file contents at the current cursor location. For each number in the file, Lotus 1–2–3 creates a number cell. For each quoted label, Lotus 1–2–3 creates a left-aligned label. Lotus 1–2–3 places successive numbers and labels from the same line in the import file in successive columns of the same row. For example, suppose a file is created with a BASIC program with the following contents:

```
"methane",   1.27,    2534711,   400
"ethane",    1.88,    1183152,   350
"propane",   2.73,    2192368,   450
```

This file can be imported to Lotus 1–2–3 in the following form:

	A	B	C	D
1	methane	1.27	2534711	400
2	ethane	1.88	1183152	350
3	propane	2.73	2192368	450

The /**File Import Number** command is very useful for viewing the contents of data files. Lotus 1–2–3 version 2 extends to row 8192, so files with as many as 8192 data points can be imported with one command.

For files with a special format of text and numbers, Lotus 1–2–3 version 2 has powerful new data-importing capabilities called *data parsing*. An ASCII file containing both labels and numbers can now be imported. Each line of the input file can be interpreted either into labels or numbers by making a template line showing where labels and numbers begin and end. The data is then placed into separate cells on the spreadsheet. This function is valuable for taking the ASCII data captured from various instruments and easily using it in Lotus.

Translate from dBase II or dBase III

Data can also be exchanged with dBase II or dBase III program files. These conversions actually create new worksheet files. These utility programs are accessible via the main Lotus access program menu.

Macros

Macros allow a number of keystrokes to be executed as a result of a single keystroke. The commands to execute a macro give the user the opportunity to create program-like applications. As an example, Figure 5.11 shows a simple macro that can be used to print multiple

AA1: 1+1
LOTUS MULTIPLE COPY PRINT MACRO   LABEL

	AA	AB	AC	AD	AE	AF	AG	AH
1		2	Counter					
2		2	Comparison				Comments	
3	\p	/RECOUNTER~					Initialize COUNTER	
4		/XNEnter Number of copies:~COMPARISON~					Prompt for # of Copies	
5		/PPR{?}~Q					Select Range to Print	
6		/XICOUNTER>=COMPARISON~/XQ					IF # needed then stop	
7		/PPAGPQ					Print selected Range	
8		{GOTO}COUNTER~{EDIT}+1~					Increment counter	
9		/XGAB6~					Goto AB6	
10								
11								
12								
13								
14								
15								
16								
17								
18								
19								
20								

Figure 5.11   Example of a simple macro used to print multiple copies of a worksheet.

copies of a worksheet. The comments on each line explain what the macro will perform. If the same keystroke sequence is used over and over, then some time can be saved by making the keystrokes into a macro.

The printing macro is very simple. Macros can be written to make a spreadsheet application entirely menu-driven from data entry to finished reports. This menu-driven ability allows Lotus to be used by a person not yet familiar with the program or the PC. A new technician can operate a data system by using a macro after being given just a few simple instructions. The macro language provided in Lotus 1–2–3 yields another dimension in automation.

Figure 5.12 shows the code for a macro. This code is an example of a completely automatic application written as a macro. This application will be automatically loaded whenever Lotus is run. Figure 5.13 shows the macro prompt for a file name. After the file name is entered, the data from that file is imported, calculations are made, and a report is displayed as shown in Figure 5.14. The report can then be printed by selecting another menu item.

G2: 'Press Alt-R to Import more Data   READY

	F	G	H	I	J	K	L	M
1		1						COMMENTS
2	YORN	Press Alt-R to Import more Data						Cell to Start Import
3	\0	{GOTO}a10~						Prompt for file
4	and \R	/FIN{?}~						Print Report?
5		/XNPrint This Report 0=NO 1=YES ?~YORN~						To Print or not ?
6		/XIYORN<=0~/XGG8~						PRINT
7		/PPAGPQ						Display Instructions
8		{GOTO}g2~						Quit
9		/XQ						
10								
11								
12								
13								
14								
15								
16								
17								
18								
19								
20								

Figure 5.12   Automatic-loading macro that prompts the user for operation. Macros can be written to automate any spreadsheet operation.

```
AlØ:
Enter name of file to import:                                              CMD MENU
REDATA   SOLDATA   MACRODAT
     A       B        C           D          E          F          G
1              Liquid Chromatograpy Data                 YORN       1
2                 Area Percent Report                     \Ø        Press Alt-R to
3                                                                    {GOTO}alØ~
4        Sample:  QA #2345    Date:      Ø4-Mar-85 and \R            /FIN{?}~
5        Column:  Sephadex    Operator:GIO                           /XNPrint This R
6        Solvent: Methanol/Water  Inst No.:LC #1234                  /XIYORN<=Ø~/XGG
7        Notes:   Report Using a MACRO to IMPORT Data                /PPAGPQ
8                                                                    {GOTO}g2~
9        Peak No. Ret. Time  Area   % Total Area                    /XQ
10                                    ERR
11                                    ERR
12                                    ERR
13                                    ERR
14                                    ERR
15                                    ERR
16                                    ERR
17                                    ERR
18                                   ------
19                             Ø      ERR
20
```

Figure 5.13   The macro prompt for the file to import.

A10: 1
Print This Report 0=NO 1=YES ?

CMD EDIT

	A	B	C	D	E	F	G
						YORN	
							1
1		Liquid Chromatograpy Data					Press Alt-R to
2		Area Percent Report				=0	{GOTO}al0~
3							/FIN{?}~
4	Sample:	QA #2345		Date:	04-Mar-85 and \R		/XNPrint This R
5	Column:	Sephadex		Operator:GIO			/XIYORN<=0~/XGG
6	Solvent:	Methanol/Water		Inst No.:LC #1234			/PPAGPQ
7	Notes:	Report Using a MACRO to IMPORT Data					{GOTO}g2~
8							/XQ
9	Peak No.	Ret. Time	Area	% Total Area			
10	1	2.34	12345	1.160			
11	2	2.67	34567	3.248			
12	3	3.68	78904	7.414			
13	4	4.98	5674	0.533			
14	5	6.89	12356	1.161			
15	6	9.78	896654	84.249			
16	7	10.76	9223	0.867			
17	8	13.45	14568	1.369			
18			————	————			
19			1064291	100.000			
20							

Figure 5.14 Report shown on the screen. The report can also be printed by selecting the print menu item.

Products Mentioned in This Chapter

Lotus 1–2–3 version 2 ($495), Lotus Development Corp., 161 First Street, Cambridge, MA 02142. (617) 577–8500

Lotus Symphony ($695), Lotus Development Corp., 161 First Street, Cambridge, MA 02142. (617) 577–8500

Framework II ($495), Ashton–Tate, 10150 West Jefferson Blvd., Culver City, CA 90230. (213) 204–5570

Javelin, ($695) Javelin Software Corporation, One Kendall Square, Building 200, Cambridge, MA 02139. (800) JAVELIN

VP-planner ($99), Paperback Software International, 2830 Ninth Street, Berkeley, CA 94710. (415) 644–2116

Chapter

Graphics

raphic images communicate ideas faster and more dramatically than text or computer printouts. For scientific and engineering applications, graphing data can provide the tools for observing trends, viewing complex interactive factors, and making the comparison of data easy. Thus, graphics use facilitates all aspects of data analysis. With an IBM PC or a compatible computer outfitted with a graphics card and graphics monitor, graphics can be created quickly, easily, and economically. By customizing a basic system with add-on monitors, graphics cards, and plotters, a graphics-oriented work station can be created for data analysis or information communication. Graphics-oriented software aids in making a computer system easy to use for novice and experienced users. The use of graphic symbols, or *icons*, rather than words can shorten the learning time for a computer application.

The graphics capabilities of the IBM PC have increased rapidly from crude, low-resolution systems to the current powerful systems taking advantage of the speed and additional computing power of the PC/AT. Some of these systems can provide 1024 × 1024 pixels of resolution with the potential of 16.7 million colors.

The future looks bright for graphics applications. Two major markets have already evolved. The scientific and engineering market, driven mainly by computer-aided design (CAD) applications, should have more than 2 million installations by the end of this decade. Meanwhile, business graphics applications are expected to be practiced on more than 10 million PC systems by the end of this decade.

A Graph Is Worth 1000 Data Points

Scientists can use graphics in two major ways. First, graphics can be used to analyze data. Graphical analysis of this sort is not new to

1000–4/87/0109$12.25/1 © 1987 American Chemical Society

scientific investigation. Historically, one of the premier applications of minicomputers in scientific laboratories was the real-time analysis of data using graphics. The IBM PC continues this aspect of computer systems, as shown in Figure 6.1. A second application for graphics is the communication of the results of analysis. When results are presented in visual format, complex data is easier to digest and analyze, and relationships become immediately evident as shown in Figure 6.2. As a tool for communication, graphics use is unparalleled. Graphics adds a high degree of professionalism and can impress, persuade, and dramatize a message quickly and effectively.

Graphics used for communicating ideas or results can also be placed in two categories: peer, or internal, applications and presentation-quality applications. Black-and-white output on plain paper is normally sufficient for internal applications, although high resolution and color are more important for presentation-quality graphics in slides and overhead transparencies.

Integrating graphics into daily data analysis can have positive effects on data quality in the laboratory. Compare the two data presentation formats shown in Figure 6.3. The graphic representation of the data will quickly get the attention of the instrument operator, who will start the procedures to solve the problem.

Caution with Graphics Applications

Graphics applications, like all other applications on PCs, are a combination of hardware and software. The huge number and wide variety of hardware and software vendors offers a mixed blessing for users of graphics applications. The variety is good because of the opportunity to build a computer system that meets the user's specific needs. On the negative side, most graphics applications have no de facto standards, unlike disk drives, tapes, and text printers; it is possible to purchase hardware and software that should work but do not produce the expected results.

I suggest the same strategy advocated for other applications. First, find the software, and then shop for the hardware necessary to execute the software.

Many advertisements and product reviews and much literature show the results of applications, but those results can only be achieved with a specific set of hardware and software. The same software package may work on other hardware and software configurations, but with less appealing results. For example, some compa-

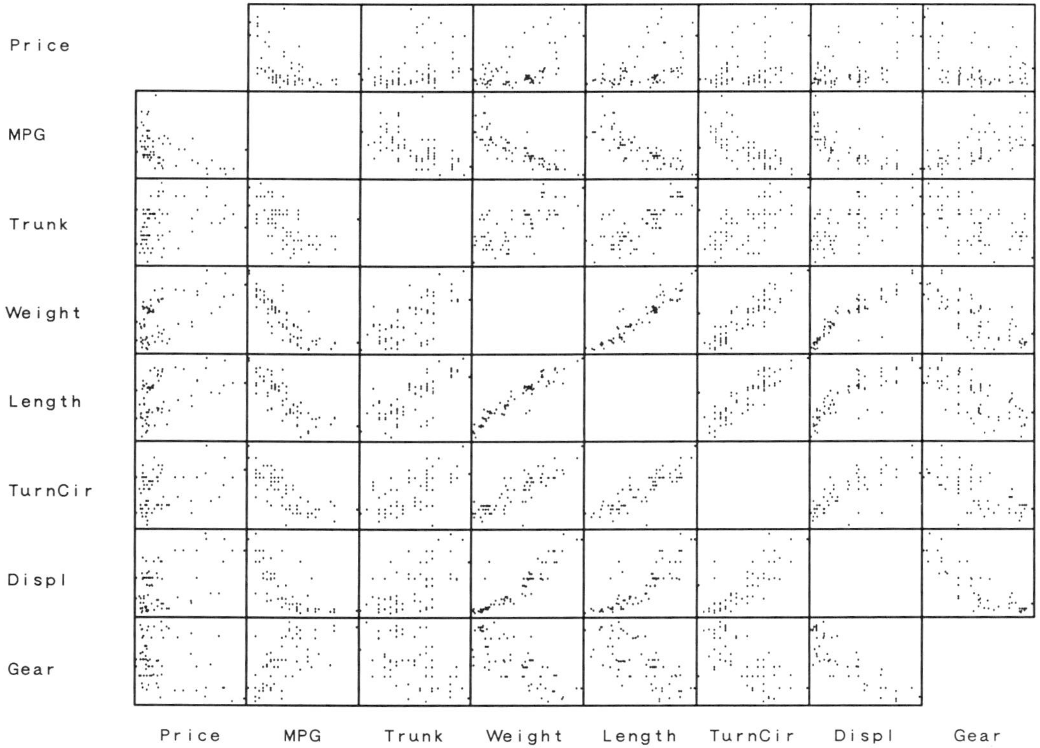

Figure 6.1 Effective use of graphics to analyze statistical data. STATA/Graphics divides the screen into 64 separate graphs that show the scatter-plot correlation of every combination of up to eight variables. This overview of data is excellent because one screen shows the variables that seem to be correlated and those that seem to be entirely independent. (Courtesy of STATA/Graphics.)

Relational Databases

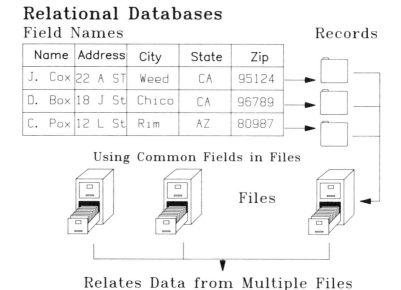

Figure 6.2 Presentation-quality graphics created by using an IBM PC and Freelance.

nies that produce graphics boards advertise color and a high-resolution mode of 640 × 400 pixels, but these same boards can only produce color in the 320 × 200 pixels mode. The only real test for any system is to see the configuration in operation with the software under consideration. Most software vendors have demonstration disks that anyone can try for specific sets of hardware.

Graphics applications are more dependent on hardware than any application discussed so far. If the right combination of monitor and graphics card for a particular software package is not available, the graphics application will not work. Similarly, having a super high-resolution, multicolor graphics board, monitor, and multipen plotter does not guarantee that the software purchased will make full, or any, use of that hardware configuration. Higher resolution graphics is just now starting to be supported by graphics-oriented software packages.

This confusion would disappear if PC software vendors would adopt IBM's support for device-independent graphics standards such as the Virtual-Device Interface (VDI) and the Graphical-Kernal System (GKS). These standards, if adopted, would allow a software vendor to write one version of a product and have the product work on many different output devices, graphics controllers, and monitors. Little or no support for these graphics standards has resulted on the part of

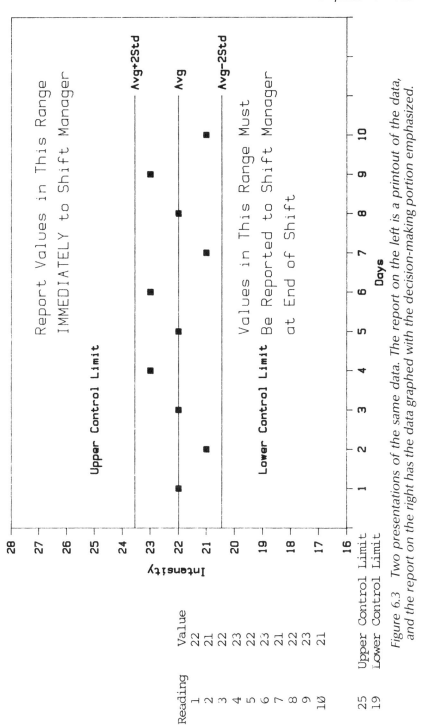

Figure 6.3 *Two presentations of the same data. The report on the left is a printout of the data, and the report on the right has the data graphed with the decision-making portion emphasized.*

application developers because of the huge speed loss inflicted on a program that gains this independence. The speed loss is due to the additional program steps that must be executed to use the standard. So the future points to a continuation of the current state of confusion, and users should exercise caution when shopping for any graphics application.

Few, if any, of the new high-resolution graphics boards have high-level language support. Until new versions of the language compilers or interpreters are developed and distributed, the only way the new boards can be used is through assembly language drivers called from high-level languages. This will be the normal mode of operation until a graphics standard is developed and widely used. But until graphics hardware development slows, which is doubtful in the next 4–6 years, graphics software will always lag behind hardware developments.

Another important word of caution regards graphics plotting. Just because a graphics program can plot a beautiful picture or diagram on a monitor does not mean a copy can easily be made of the work. Most graphics can be sent to a dot-matrix printer, but for a program to plot data, the program must have special additional programming to drive a plotter. The software required to create a plot on a plotter is completely different from the software required to generate a plot on a graphics monitor. The plotter software must send a series of commands to the plotter and instruct the plotter where and how to draw lines on the paper. These commands provide sets of locations to start and end a specific vector, or line, on the plotter. The software needed to plot a graph on the monitor sends commands that control which screen dots are to be illuminated and in which color. This software is *screen-dot-* or *pixel-based*. Plotting on a plotter requires *object-* or *vector-based* software.

Graphics Software for the IBM PC

Graphics software for the IBM PC and PC/AT covers the spectrum from simple data plotting and screen image-making programs to complete CAD and high-quality graphic design systems. These programs fall into two categories. *Drawing programs* allow graphs and charts to be generated from data; the graph types are dictated by software. *Painting programs* allow free manipulation of each screen pixel to generate images. Drawing programs are most useful for the analysis and communication of scientific data, and painting programs are needed by creative scientific communicators.

Drawing Programs

Drawing graphics software provides the ability to create many types of graphs. These graphs are useful for communicating results and analyzing data. The graphics software provided in three software packages will be examined: Lotus 1–2–3, which provides graphics integrated with its spreadsheet analysis system; Graphwriter 4.2, a stand-alone graphics program providing a larger selection of graph types; and the freehand drawing program AutoCAD, which is a CAD package. These programs are representative of the more than 70 vendors of graphics software who provide more than 90 graphics programs.

One of the most powerful features of Lotus 1–2–3 is the ability to manipulate data on the spreadsheet and then, with just a few keystrokes, provide a graph of that data on the computer screen. The graphics are generated quickly; interactive changes in the way data is manipulated can almost instantly be seen graphically. These graphs can be printed by using a compatible printer or plotted by using a compatible plotter.

Lotus 1–2–3 provides the ability to produce the basic types of line, bar, and pie graphs. If other types of graphs are needed, stand-alone graphics packages such as Graphwriter should be considered. Most stand-alone packages use data from the Lotus spreadsheet, so the user doesn't have to reenter data.

The simplest but most widely used graph is a line graph. Lotus allows up to six line graphs to be plotted at a time. These graphs can be scaled to any size. Some applications of line graphs will be discussed.

Step-by-Step Plotting: Inductively Coupled Plasma Spectroscopic Scan Plot. If this example is to be followed, Lotus 1–2–3 and a compatible graphics monitor and card must be used. Check the Lotus manual to be sure the necessary hardware can perform the graphics on the system.

Many commercial instruments have the ability to send data to another computer (*see* Section 3, which describes instrument–computer interfacing). When data is captured, it can be imported into Lotus 1–2–3 and graphically displayed. The Leeman Labs PlasmaSpec has an RS–232C interface and can send data to an IBM PC. The data for a specific inductively coupled plasma (ICP) spectroscopic scan is sent by using a format, as shown in Figure 6.4. This data is captured

"$PHG1	1	1474	11"
"$PHG1	2	1732	ØF"
"$PHG1	3	2116	ØD"
"$PHG1	4	2628	16"
"$PHG1	5	3364	15"
"$PHG1	6	4528	19"
"$PHG1	7	6636	1C"
"$PHG1	8	10896	30"
"$PHG1	9	19632	2E"
"$PHG1	1Ø	36464	38"
"$PHG1	11	75264	3A"
"$PHG1	12	152256	48"
"$PHG1	13	257664	52"
"$PHG1	14	33344Ø	46"
"$PHG1	15	353344	4C"
"$PHG1	16	3ØØ16Ø	41"
"$PHG1	17	2Ø4464	4C"
"$PHG1	18	11456Ø	4A"
"$PHG1	19	58048	43"
"$PHG1	2Ø	33392	36"
"$PHG1	21	22448	37"
"$PHG1	22	16416	36"
"$PHG1	23	13Ø4Ø	2D"
"$PHG1	24	1Ø356	35"
"$PHG1	25	852Ø	26"
"$PHG1	26	7216	28"
"$PHG1	27	6244	29"
"$PHG1	28	5392	2D"
"$PHG1	29	4744	2E"
"$PHG1	3Ø	418Ø	2Ø"
"$PHG1	31	3756	29"
"$PHG1	32	3344	23"

Figure 6.4   Format of the data sent from the ICP instrument. Each intensity value includes the letter P followed by the data point number and then the intensity.

and then loaded into Lotus 1–2–3 by using the file import numbers command (see Chapter 5). This valuable Lotus command reads text characters from a file and converts all text numbers into actual numbers in the spreadsheet. The numbers are then analyzed as usual numbers. The data can then be plotted.

Plotting the data requires a series of Lotus graph commands. The data looks like Figure 6.5. Select /**Graph** to enter the graph commands. Now select the data to plot as the cells A1–A32 by entering the commands **A** and **A1..A32**. Then select the graph type with the command **Type Line**.

	A	B	C	D
1	1474			
2	1732			
3	2116			
4	2628			
5	3364			
6	4528			
7	6636			
8	10896			
9	19632			
10	36464			
11	75264			
12	152256			
13	257664			
14	333440			
15	353344			
16	300160			
17	204464			
18	114560			
19	58048			
20	33392			

A40:

	A	B	C	D
21	22448			
22	16416			
23	13040			
24	10356			
25	8520			
26	7216			
27	6244			
28	5392			
29	4744			
30	4180			
31	3756			
32	3344			
33				
34				
35				
36				
37				

Figure 6.5   Imported data in the Lotus spreadsheet.

By selecting **View**, the work done so far can be seen. The screen should look like Figure 6.6. This crude graph did not take much time to generate. Titles and axis labels can be added in the following manner.

Select the options titles commands. Now enter under First title, **ICP Data**; under Second title, **Hg Standard Data**; under Y-axis, **Intensity**; and under X-axis, **nm**.

The display should look like Figure 6.7. This graph can be saved for printing or plotting by selecting **Save** and entering a file name. The data for the graph will be stored with the file name supplied with the extension PIC. The PIC files "hold" a Lotus graph or picture. These PIC files are then read by the print graph program or other programs such as Freelance so data can be printed or plotted.

The scale of the plot can be changed to view the data more closely. Select options scale. Currently, the option should be automatic scaling, so select manual and then select the lower and upper limits. Enter **50000** for the upper limit and **0** for the lower limit. The graph should now look like Figure 6.8.

Two more scans of data can be plotted (up to six can be plotted with Lotus 1–2–3) by *importing* the data into columns B and C. Data is imported by starting from the current position of the cursor. This allows import of a number of blocks of data into the same spreadsheet without losing the current contents and allows the new data to

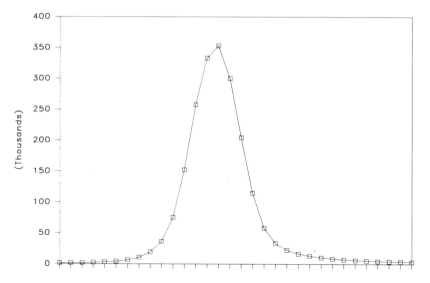

Figure 6.6   First graph of the ICP data.

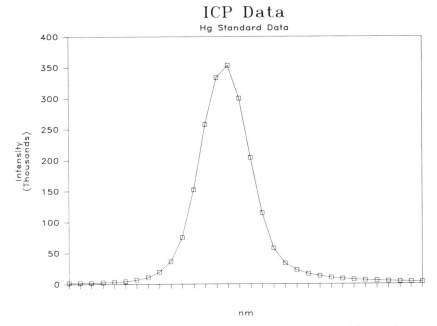

Figure 6.7   The graph of the ICP data now contains titles and labels.

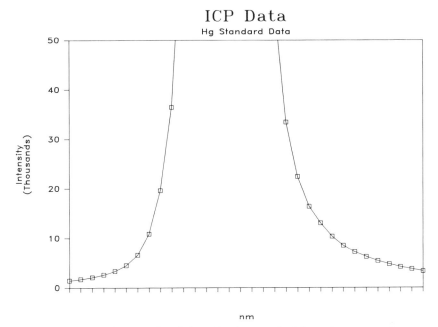

Figure 6.8   The graph of the ICP data scaled by using manual values rather than the programs default scale.

be placed wherever it is wanted. After importing two more scans of data, the spreadsheet should look like Figure 6.9.

Now this data can be plotted by defining the B and C graph-plotting ranges. The plotting ranges can be any range of numbers on the spreadsheet. Only by coincidence in this example are the A, B,

	A	B	C	D
1	1474	1179	884	
2	1732	1385	1039	
3	2116	1692	1269	
4	2628	2102	1576	
5	3364	2691	2018	
6	4528	3622	2716	
7	6636	5308	3981	
8	10896	8716	6537	
9	19632	15705	11779	
10	36464	29171	21878	
11	75264	60211	45158	
12	152256	121804	91353	
13	257664	206131	154598	
14	333440	266752	200064	
15	353344	282675	212006	
16	300160	240128	180096	
17	204464	163571	122678	
18	114560	91648	68736	
19	58048	46438	34828	
20	33392	26713	20035	

	A	B	C	D
21	22448	17958	13468	
22	16416	13132	9849	
23	13040	10432	7824	
24	10356	8284	6213	
25	8520	6816	5112	
26	7216	5772	4329	
27	6244	4995	3746	
28	5392	4313	3235	
29	4744	3795	2846	
30	4180	3344	2508	
31	3756	3004	2253	
32	3344	2675	2006	
33				
34				
35				
36				
37				

Figure 6.9 Spreadsheet with multiple scans of ICP data.

and C graph ranges in columns A, B, and C of the spreadsheet. Select the B and C graph ranges by entering the commands /**Graph B B1..B32 C C1..C32**.

Now the graph should look like Figure 6.10. A legend can be added to the graph by using options legend. Enter for Legend A: **Sample**, for Legend B: **High Standard**, and for Legend C: **Low Standard**. Escape or use the quit command to move back to the graph command level and view the plot, which should look like Figure 6.11. Any set of data can be plotted by using these techniques. Plotting multiple scans of data allows easy data comparison.

Interactive Plotting: Least-Squares Fit Calibration Curve. In Chapter 5, a Lotus spreadsheet was generated to calculate the linear least-squares fit to a set of calibration data. By using the same spreadsheet calculations, this data can be plotted. Start with the calculations as described in Chapter 4 and shown in Figure 6.12. The least-squares fit line needs to be plotted, and each data point should be displayed. To do these steps, two plot ranges must be defined, and different display formats must be used for each. The *default method* of line graph display is a combination of lines joining each data point

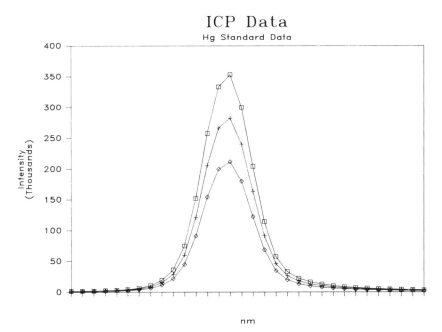

Figure 6.10   Graph of multiple ICP scans.

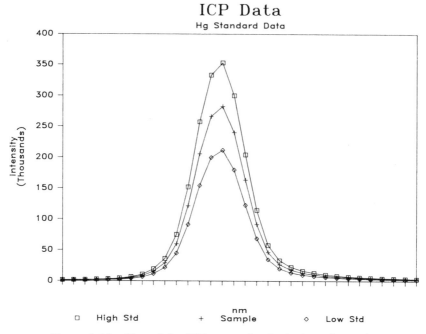

Figure 6.11   Plot of the ICP scans that includes a legend.

	A	B	C	D
1	Calibration Data For Methane			
2				
3	Concentration	Area/1000		
4				
5	0	106.5		106.5
6	100	210	21000	205
7	200	304	60800	303.5
8	300	397	119100	402
9	400	489	195600	500.5
10	500	610	305000	599
11	600	697.5		697.5
12				
13	Slope of Line:	0.985		
14	Y-Axis Intercep	106.5		
15				
16				
17				
18				
19				
20				

Figure 6.12   Least-squares fit calculation spreadsheet.

and a symbol displayed for each data point. The way each range of data will be plotted can be changed. For this example, one range will be the least-squares data joined by lines but without symbols. The second range will be the calibration data points with symbols only.

To implement this graph, do the following: Create a new column of data that will hold the calculated least-squares fit data. This column contains calculations of the equation for a line with the computed least-squares fit coefficients substituted. This column will become the A graph range.

Now identify the B range as the calibration data points. View the data; the display should look like Figure 6.13. Notice that in the default display mode, each set of data has lines and symbols. Now the graph format can be changed for each range. For the A graph range, enter **Options Format A Lines**. Then enter the B range format as symbols only with the command **Format B Symbols**. After these changes, the graph should look like Figure 6.14.

Whenever a calibration value is changed, the effect can instantly be seen on the graph. As an example, change the intensity of level 2 to **50000**. Now view the new least-squares fit line by simply pressing F10, the graph key. This key will display the last displayed graph.

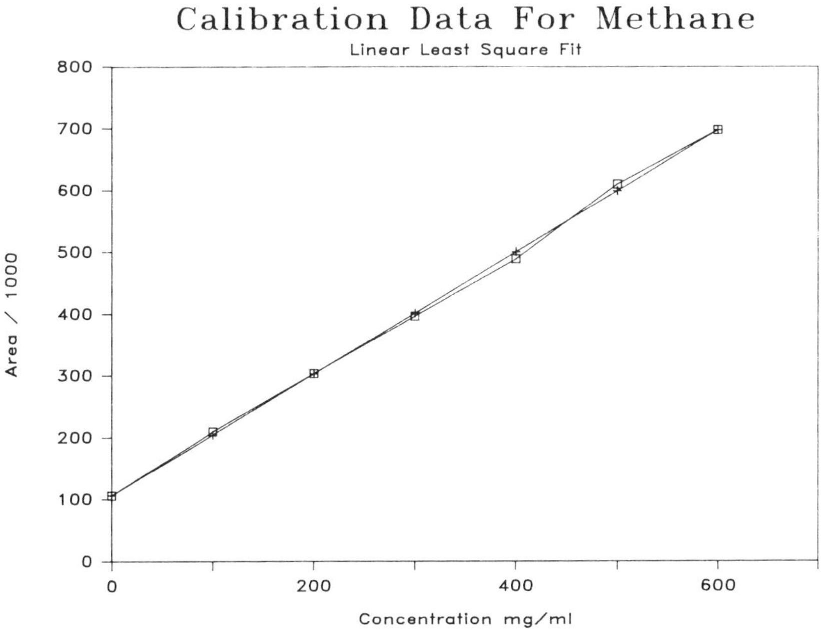

Figure 6.13   Calibration data plot.

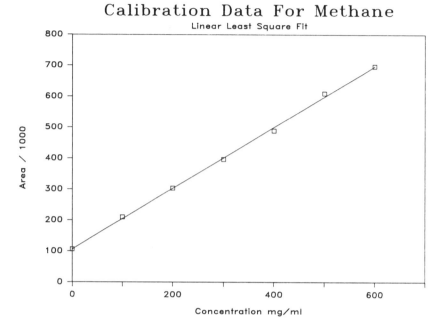

Figure 6.14   Graph showing effective use of a combination of symbols and lines.

Thus, a linear least-squares calibration application with graphics has been created that can be used for almost any application.

Managing Multiple Graphs: Quality Data Charts.   In Chapter 5, a quality data spreadsheet was started. Built-in functions were used to compute the average and standard deviation of a set of experimental readings. Now this data can be plotted and the results viewed. A number of plot settings can also be set up and saved, so switching from one view of the data to another is possible.

Start with the following spreadsheet data: in A3, **100**; A4, **110**; A5, **120**; A6, **105**; A7, **130**; A8, **112**; A9, **111**; A10, **109**; A11, **108**; A12, **108**; A13, **110**; A14, **111**; A15, **109**; A16, **110**; A17, **109**; and A18, **108**. To plot the data with the calculated values, three new ranges must be created: a range with the average, a range with the average plus one standard deviation, and a range with the average minus one standard deviation. To make the plots, a range of cells containing the average and standard deviation data points must be defined. A number of cells will contain the same values. This data can be set up by using the Lotus features called the cell absolute reference and named ranges.

When the Lotus copy command is invoked to copy the contents of a cell or range of cells to other cells on the spreadsheet, the program automatically updates cell references within the copied cells. For example, if columns of numbers are summed with the command +@SUM(A1..A30), and cell contents are copied to columns B, C, and D, then the cell references are automatically changed to +@SUM(B1..B30), +@SUM(C1..C30), and +@SUM(D1..D30), respectively. This updating of references is the default method used by Lotus for all copy commands. Some cases require use of the copy command, but the cell references should not be updated. When this situation occurs, an absolute reference is needed as a cell identifier.

To identify a cell identifier as an absolute reference, a dollar sign symbol, **$**, is placed in front of one or both of the cell column and row identifiers. For example, the absolute reference for cell B4 would be **$B$4**. Similarly, an absolute reference to a range of cells requires **$A$1..$A$30**, and a named range requires a **$** in front of the range name. When a cell containing an absolute reference is copied, the cell identifier will not be updated and will stay the same. This copying is needed for the quality control chart application.

A range of cells must now be created that contains the same value; the easiest way to do this is to make absolute references to specific cells and then to copy the cell contents to the cells in the range. In the current data, the values are listed in the A column. First, a named range can be created for the range of cells to be tracked. The command / **Range Name Create DATA A3..A13** will accomplish this.

The average of the data can then be computed and displayed in cell B3 by entering +@AVG($DATA). This function can be copied to the other cells in the B column with the command / **Copy B3..B3 to B4..B13**. Now the upper and lower standard deviation line data can be constructed with the functions +@AVG($DATA)+@STD($DATA) in cell C3 and +@AVG($DATA)-@STD($DATA) in cell D3. Now this data can be copied with the command / **Copy C3..D3 to C4..D13**.

Now four columns of data have to be graphed. First, assign the ranges to be plotted. Assign the A plot range by using the data-named range with the command / **Graph A DATA**. Now assign the B, C, and D ranges with the command **B B3..B13 C C3..C13 D D3..D13**. Select line graphs with the command **Type Line**. The display should look like Figure 6.15. The graph still needs more work to look professional.

First remove the symbols from the average and standard deviation limit lines and the lines between the data values. Do this by using the format command under graph options. Enter the commands **Options**

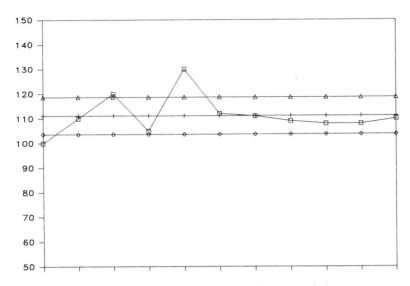

Figure 6.15   First plot of the quality control chart.

Format A Symbols B Lines C Lines D Lines Quit Quit. Now view the work. The display should look like Figure 6.16.

Now, a title, label, and legend can be added for each of the range of plotted cells. Enter the following commands:

Options Titles First Quality Control Chart
Options Titles Second May Week 2
Options Titles X-Axis Days
Options Titles Y-Axis Intensity
Options Legend A Data B Average C Avg+1Std D Avg-1Std

Now view the work, which should look like Figure 6.17. This chart is very acceptable. It can be improved and will be, but first data can be added to the plot. This addition is where the use of named ranges will pay off. Add five more points to the data in cells A14–A18. To include these points into the calculations, simply redefine the data range with the command / **Range Name Create Data A3..A18**. All of the calculations will now be performed on this new range of cells. Plot the data by copying the average and standard deviation data to more cells and redefining the cell ranges for the B, C, and D plots. The command for copying the data is / **Copy B3..D3 to B14..D18**. The command for redefining the plot ranges is / **Graph B B3..B18 C C3..C18 D D3..D18**. Now view the work; it should look like Figure 6.18.

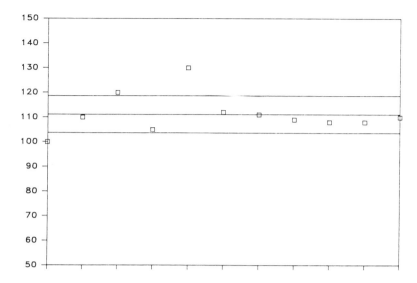

Figure 6.16   Effective use of symbols and lines with the quality data chart.

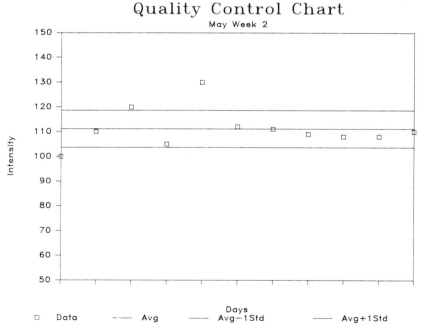

Figure 6.17   Quality data chart with titles and labels.

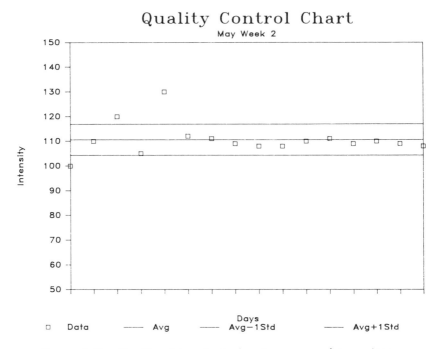

Figure 6.18   Quality data chart showing more data points, which can be added easily.

To give the graph more appeal, the plot scales and ranges can be altered. The default scale places the highest value at or near the top of the plot and the lowest values at or near the bottom of the graph. The scale can be changed to manual mode where the upper and lower values can be selected. To change the Y-scale, enter /**Graph Options Scale Y-Scale Manual Lower 0 Upper 20 Quit**. Now view the result as shown in Figure 6.19. The goal is to have the data points on the X-axis not plotted on the edges of the graph yet not have the average and standard deviation lines extend to the edge. To achieve these conditions, the plot ranges must be redefined to include a blank cell before and after the range of data, and the average and standard deviation plots must include two additional cells. This step can be accomplished by copying the contents of the line cells to rows 2 and 19 by entering /**Copy B3..D3 to B2..D2** and /**Copy B3..D3 to B19..D19**, respectively. Now redefine the plot ranges with the commands /**Graph A A2..A19 B B2..B19 C C2..C19 D D2..D19**. Now view the work, which should look like Figure 6.20.

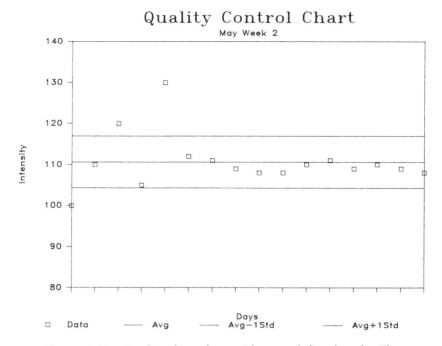

Figure 6.19   Quality data chart with user-defined scale. The range of the Y-scale has been changed.

Sometimes it is more visually effective to have labels on the graphed lines rather than at the bottom of the graph. This can also be done with Lotus graphics by using the data labels command. First, more room must be made on the right side of the graph for the labels. Do this by including a number of blank cells in outplot ranges with the command **/Graph A A2..A25 B B2..B25 C C2..C25 D D2..D25**. A data label can now be attached to the last data value in each set of data. The plot should look like Figure 6.21.

Printing and Plotting Graphs.   Lotus 1–2–3 provides the ability to print or plot graphs on printers or X–Y plotters by using a separate program called Print Graph. Lotus Print Graph supports a wide range of output devices, so copies of graphs on a dot-matrix printer or presentation-quality copies on overhead transparencies using an X–Y plotter can be obtained. To print or plot any graphics, first save the current graph with the save command within the graph commands. Making a copy of the graph screen with standard Lotus software is not

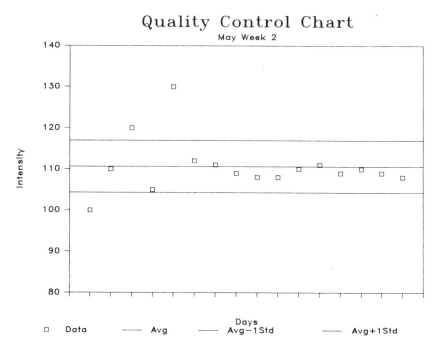

Figure 6.20   Plot using blanks to make a more pleasing display. This technique also moves the first and last data points off the graph edge.

possible. If a quick copy on a dot-matrix printer is needed, contact DOMUS Software. DOMUS provides a utility program for printing the graphics screen when the PrtSc key is pressed.

A majority of the graphs in this book were generated on an EPSON FX–80 dot-matrix printer. Lotus provides four density levels for the EPSON FX–80. Each level becomes more dense and easier to look at. Each successive level also takes longer to print.

Graphs can also be plotted on compatible plotters. If a compatible plotter with multiple pens is used, plots can be automatically done in multiple colors. Be sure to save the graph with the color command activated under the Graph Options menu if different colors are desired. If a single-pen plotter is used, different colors can be obtained by commanding the Print Graph program to pause after each range of data is plotted. This pause allows the user to change the plot pen.

For a quick copy of the graphs seen on the monitor, purchase PRINT SCREEN from DOMUS Software Limited. This add-on program

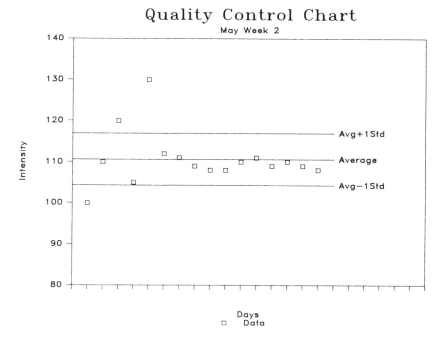

Figure 6.21   Quality data chart using labels on the graphed lines. A more effective presentation of the data results.

is one of a multitude for Lotus 1–2–3. This program will print the Lotus graph that appears on the screen immediately on a dot-matrix printer. The quality is not as good as that produced with the Print Graph program, but if an immediate print is needed, PRINT SCREEN is an excellent compromise.

Lotus also allows data to be plotted by using bar, stacked bar, pie, and X–Y graphs. Examples of these charts are shown in Figures 6.22–6.25. The strategies used with line charts can also be extended to these graphs.

Graphwriter 4.2, beyond Simple Graphs

Lotus 1–2–3 allows a limited set of graphs to be created. Other programs provide a wider range of graphs, such as Graphwriter 4.2 from Graphics Communications. Graphwriter provides many more graph types not available in Lotus 1–2–3 (Figures 6.26–6.33). The program will read data from Lotus files, or the user can enter data directly into the program.

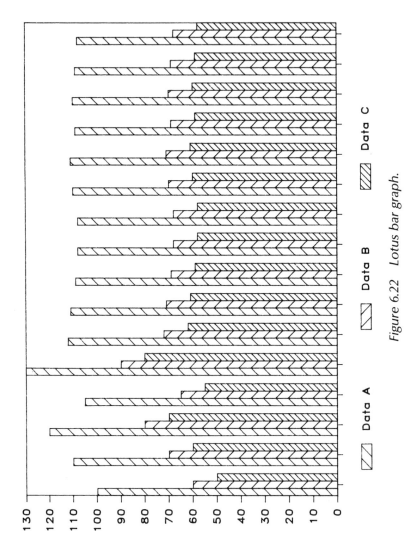

Figure 6.22 Lotus bar graph.

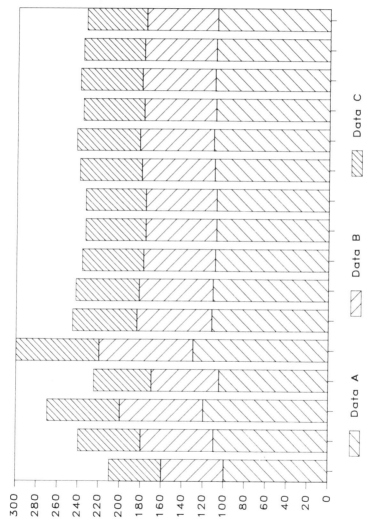

Figure 6.23 Lotus stacked bar graph.

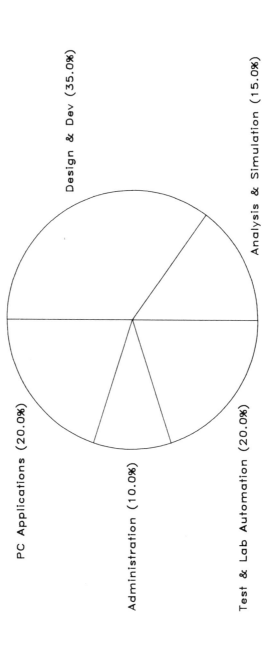

Figure 6.24 Lotus pie chart. (Courtesy of IBM.)

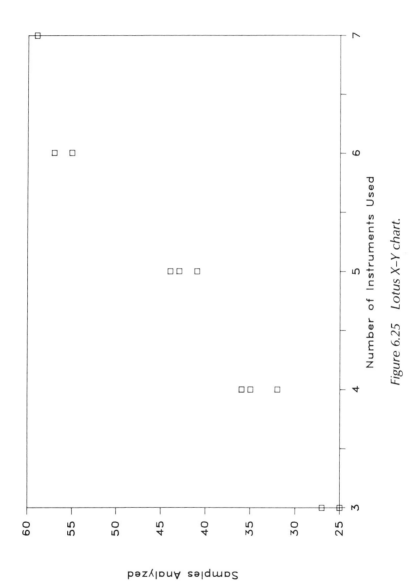

Figure 6.25 Lotus X–Y chart.

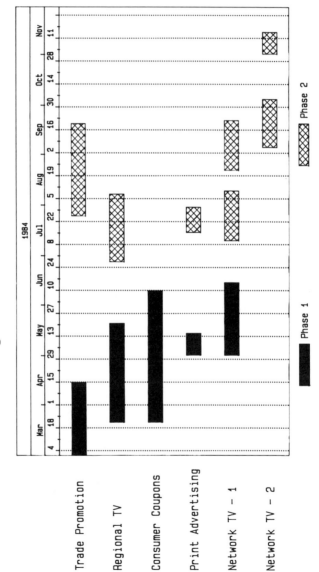

Figure 6.26 Gantt charts (see discussion in Chapter 8).

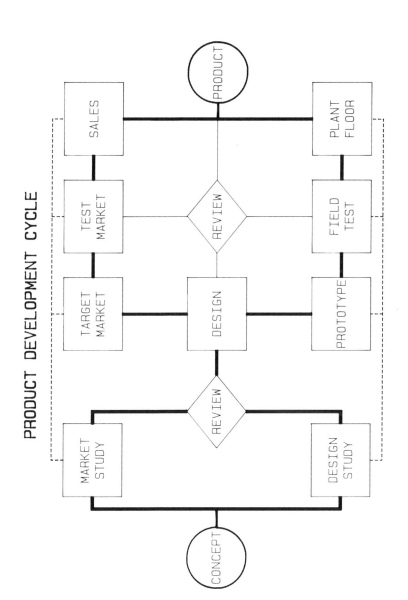

Figure 6.27 Organizational charts. (Courtesy of Graphic Communications.)

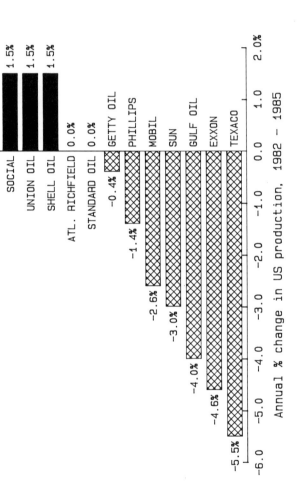

Figure 6.28 Horizontal bar charts. (Courtesy of Graphic Communications.)

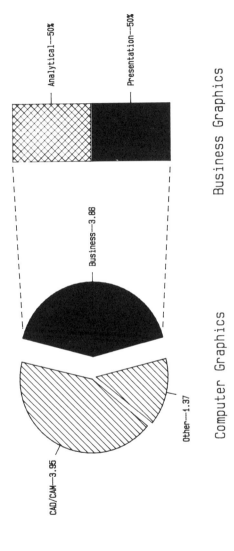

Figure 6.29   Paired bar charts. (Courtesy of Graphic Communications.)

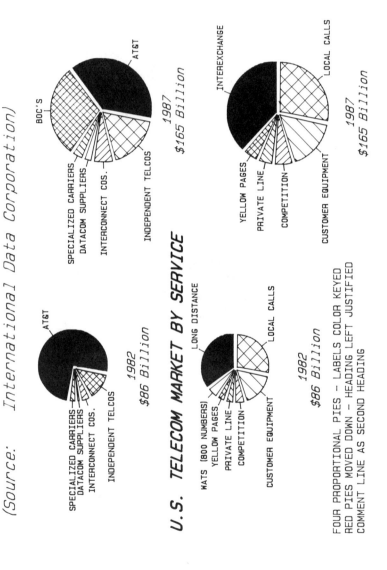

Figure 6.30 Scatter charts. (Courtesy of Graphic Communications.)

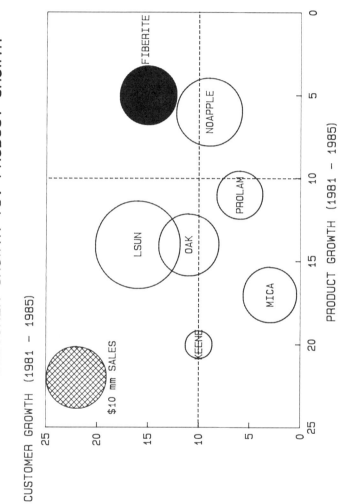

Figure 6.31   Bubble charts. (Courtesy of Graphic Communications.)

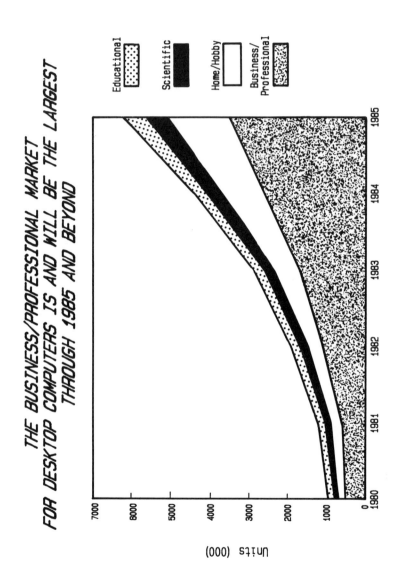

Figure 6.32 Surface charts. The patterns shown here in black and white represent different colors. (Courtesy of Graphic Communications.)

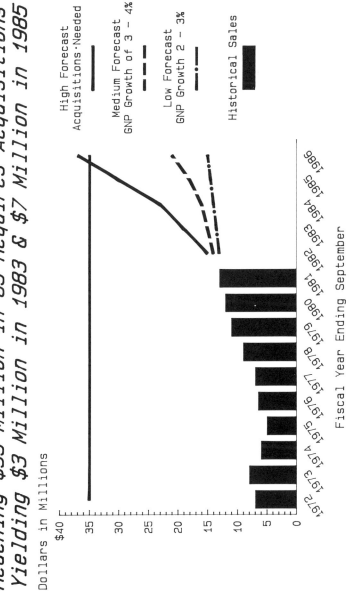

Figure 6.33 *Overlay charts. The dotted and dashed lines shown here in black and white represent different colors. (Courtesy of Graphic Communications.)*

PC-Based Computer-Aided Design

CAD software permits designers to move, copy, scale, and rotate any defined object. The PC versions of these systems support most features found on mainframe- or minicomputer-based systems including rubberbanding, zoom magnification, and wireframe three-dimensional modeling. With improved PC hardware such as the PC/AT with faster processing speeds and better graphics displays and cards, these packages are proving cost-effective and powerful enough for engineering design.

One of the most popular PC-based CAD programs is AutoCAD 2. The current version of this program offers two drafting modules and an extension that does limited three-dimensional design. AutoCAD supports a number of data input devices, including mice and digitizing pads. AutoCAD supports output to a wide variety of plotters of all sizes, has the ability to share files with a number of mainframe CAD systems, and can save drawing attributes in a separate data base. The system has a macro language that allows experienced users to set up drawings quickly by using specific modification routines. Figure 6.34 shows a typical drawing created by AutoCAD.

Painting Programs

The drawing programs just discussed are characterized by displaying data in various ways and formats. Painting programs are distinguished

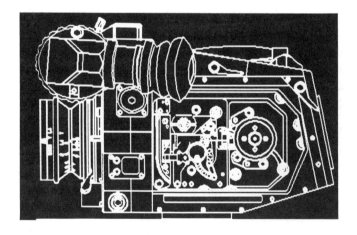

Figure 6.34 Plans created by using AutoCAD. (Courtesy of AutoDesk.)

by how easily images can be created and manipulated on the computer screen. Rather than wading through multilevel menus, these programs rely on the user's own drawing or hand movements to create the graphics. A stroke of a hand will create a stroke of color and forms on the screen. The programs are best used with a mechanical pointing device such as a mouse, joy stick, or digitizing pad.

PC Paint. PC Paint is one of many programs that allow original images to be created or diagrams to be captured and enhanced from other software programs. This program is most easily used with a hand-held mouse, which becomes an on-screen paintbrush. The program can manipulate shapes and lines on the screen. On-screen icons and pop-up menus make it easy to choose options. Options include image duplicating, rotating, erasing, and magnifying. Texture and colors can also be added and edited, including "spray paint" patterns.

Such programs are excellent for adding interest to on-screen images and for producing on-screen slide shows. These programs fall short in output. What is seen on the computer screen is the best image to be expected on paper. To get presentation-quality graphics from on-screen images, a CAD system of a program, like Freelance, must be used.

Freelance, the Hybrid Champion. The numerous painting programs like PC Paint permit memory locations, which dictate the image on the screen, to be manipulated. Each of the pixels, or pels, on the screen can be controlled to create the desired picture or image. This control allows the user to make great pictures on a PC display and provides only a *screen dump*, or one-to-one reproduction of the screen, when printed. The resulting output resolution is not presentation-quality.

The program Freelance stores its images as a series of objects or vectors rather than by just manipulating the pixels on the screen. This form of image storage is more difficult; however, it pays excellent dividends when images are sent to a plotter. Rather than medium- or low-resolution images, the plotter can be instructed to produce high-resolution, presentation-quality output.

Freelance is thus a hybrid product capable of allowing the user to create and display interesting images on a PC screen and generate high-resolution output similar to that found in CAD applications.

Other than user-defined images, the main input for Freelance is Lotus 1-2-3 graphics files. These PIC files can be read into Freelance and then manipulated.

Libraries of common symbols can be stored and recalled when necessary. BREGO Research has created a library of chemical symbols that includes more than 200 chemical structures. These structures can be recalled, edited, and used to create professional presentations in minutes. For example, Figure 6.35 shows a Lotus 1-2-3 graph of a chromatogram. The Lotus graph can be read into Freelance and enhanced to produce the graphic image shown in Figure 6.36. Chemical structures were recalled from a library of structures (Figure 6.37). All of the enhancements, including text tables, text font, and size changes, were added by using Freelance.

Graphics Monitors and Boards for the IBM PC

Hardware available for graphics on an IBM PC is available from many sources. The main components are monitors, graphics boards, printers, and plotters. These systems have continued to make vast improvements over the past few years. Each new system provides better resolution, more colors, and more features.

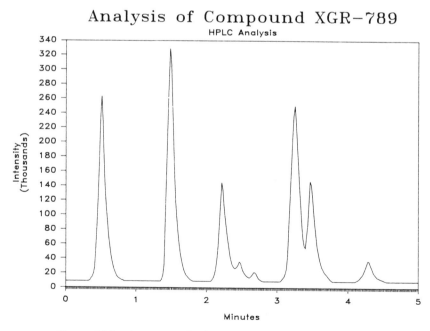

Figure 6.35 Lotus graph showing chromatographic data.

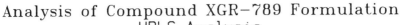

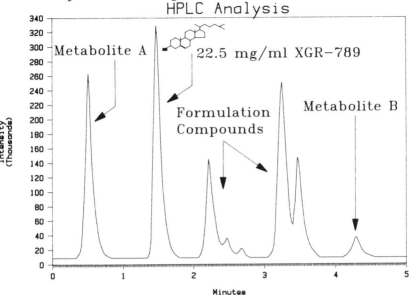

*Figure 6.36   Obviously improved graph after enhancement
using Freelance. Structure was assembled by using symbols
from the Freelance library of chemical structures.*

Surveying the graphics monitors and boards available for the IBM
PC first requires a historical look at the development of graphics on
the IBM PC. The first IBM PCs had three monitors and two boards for
a computer display from which to choose. The choices were a
monochrome board and monitor with no graphics capability, or a
color–graphics card that could drive a television set via composite
signal or drive red, green, and blue (RGB) monitors to display up to
four colors at a time from a set of either 16 colors in medium-
resolution mode (320 × 200 pixels), or black and a choice of one other
color in high-resolution mode (640 × 200 pixels). None of these early
systems would excite a computer graphics expert, but the availability
of even the most basic graphics on the IBM PC, plus the obvious
expandability of the computer, stimulated the development of better
graphics systems.

Because IBM was slow to provide higher resolution graphics
systems, a chaotic battle was fought by a number of independent
board and monitor vendors to fill the void. Today, more than 20

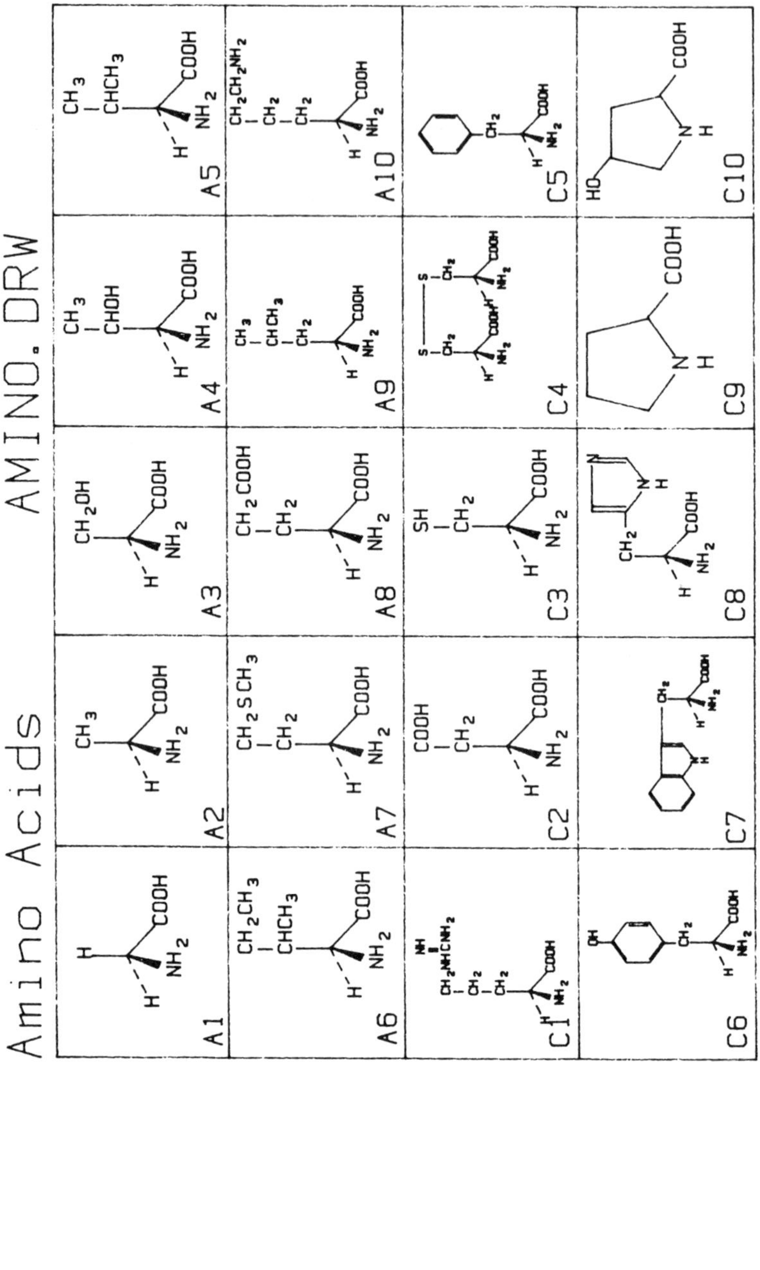

Figure 6.37 Diagram from the Freelance library of chemical structures.

companies provide display boards and monitors for the IBM PC and PC/AT. Recently, even IBM was added to the list of high-resolution graphics systems vendors when the IBM Enhanced Graphics Adapter (EGA) and monitor (640 × 350 pixels with 16 colors) and the IBM Professional Graphics Controller and monitor (640 × 480 pixels with 256 colors) were introduced.

Four groups of graphics boards and monitors have evolved to support the IBM PC. Monochrome graphics with 640 × 348 pixels and one color is the first group. Medium-resolution systems mimic the initial color–graphics systems, with 640 × 200 pixels in one color on a black background or four colors with 320 × 200 pixels resolution (Figure 6.38) in the second group. The third group has 640 × 200 or 640 × 400 pixels (Figures 6.39 and 6.40) with as many as 16 colors. Finally, for serious graphics applications, a fourth group of systems with as many as 1024 × 1024 pixels and 16.7 million colors is available.

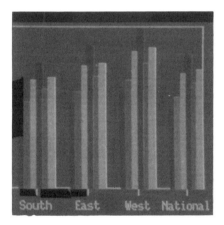

Figure 6.38 Color graphics monitor and adapter screen, shown here in black and white. (Courtesy of Paradise Systems.)

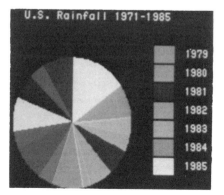

Figure 6.39 Plantronics graphics mode, shown here in black and white. (Courtesy of Paradise Systems.)

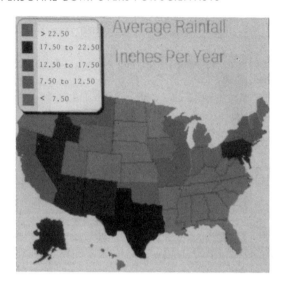

Figure 6.40 High-resolution graphics screen, shown here in black and white, using an Enhanced Graphics Adapter and Enhanced Monitor. (Courtesy of Paradise Systems.)

Hercules, the Monochrome De Facto Standard

One of the earliest improved graphics systems was provided by Hercules Computer Technology. The founder of Hercules, who purchased an IBM PC with a monochrome display, was extremely disappointed to find that the system could do no graphics. He quickly recognized his own problem as an opportunity. If he was disappointed in the nongraphics ability of the monochrome display and card, others must also have the same frustrations. So while other companies concentrated on producing the more conventional higher resolution color systems, Hercules created the Hercules card. This card could drive the normal IBM monochrome monitor to perform graphics with good resolution (720 × 348 pixels) (Figure 6.41).

The Hercules card, like all hardware, represented a useful application of electrical engineering. Unfortunately, no software could take advantage of this new technology. The Hercules card got its real boost when a newly released program, Lotus 1-2-3, could use the card by simply configuring the software with a different set of software device drivers. Revision 1.1 of Lotus 1-2-3 supported the Hercules card; thousands of Hercules cards were then sold. Today,

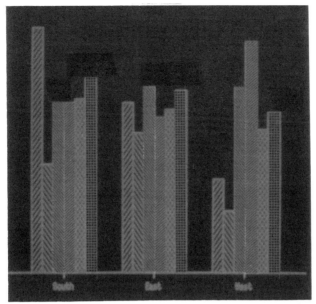

Figure 6.41 Example of Hercules monochrome graphics, shown here in black and white. (Courtesy of Paradise Systems.)

the Hercules card, because of its performance of monochrome graphics, has become the de facto standard. This combination is supported by hundreds of graphics-oriented programs, including Lotus Symphony, Ashton–Tate's Framework, the Microsoft Flight Simulator, AutoCAD, and Nelson Analytical's chromatographic data systems.

Several other vendors now offer monochrome graphics boards similar to the Hercules card, including AST Research, Sigma Designs, and Tseng Labs.

Medium-Resolution Graphics

Graphics boards and monitors with medium resolution are now provided by a number of vendors. This resolution is the most popular configuration and thus has the most software support. This resolution is good for reporting information in bar, line, and pie charts and is the favorite configuration for performing business graphics. Many scientific applications can also use this resolution, especially if the graphics software supports a plotter for output. For on-screen scientific data analysis, the highest level of graphics systems is the best, but many fewer software packages support the highest resolution systems.

Middle-of-the-Road Resolution

With IBM's introduction of the Enhanced Graphics Adapter, middle-of-the-road resolution should receive great attention from software developers. This resolution is more than adequate for most scientific applications that require on-screen data analysis. Another measure of graphic sharpness is the text character size produced by the board. Most boards produce a character made from a 5 × 7 matrix of dots within an overall space, or "box", of 8 × 8 dots as shown in Figure 6.6. For sharper characters, some vendors provide a higher number of dots. For example, the Sigma Design Dazzler and Dazzler 2 boards use a 5 × 7 character inside an 8 × 16 box.

IBM, Plantronics, Quadram, Sigma Design, and Tseng Labs also provide products for this resolution.

High-Resolution Graphics

High-resolution graphics starts at 640 × 480 pixels with 64–512 colors. These systems can have as many as 1024 × 1024 pixels with millions of colors. Most of the systems with this resolution are currently being used for CAD. Hardware of this quality can easily support most other scientific applications. None of these systems have the application software capable of using the available hardware. In the next few years, a number of excellent graphics-oriented packages for scientific data analysis will be developed that use this hardware.

Output Devices: Printers and Plotters

Most graphics software packages support a wide variety of printers and plotters. Dot-matrix printers can be used to produce medium-resolution copies of graphs. These plots can be used for daily use and data analysis. Not all dot-matrix printers can print graphics. If a printer is to be used for output, the software must support the available printer. The graphics used in the EPSON printers, early models (MX–80) of which were sold by IBM, has become a de facto standard for dot-matrix graphics. Most software programs that produce dot-matrix graphics will support the EPSON graphics.

The finest quality graphics requires a plotter. Plotters compatible with the IBM PC range in price from $400 to $6000. When purchasing a plotter, be absolutely sure the desired software supports the model under consideration. Unlike printers, which have a standard text-printing format allowing text to be printed from almost any program, a plotter must be driven by a specific set of instructions. To meet the needs for plotter instructions, the manufacturer will build in a graphics language. Although no standard language for plotters has emerged, Hewlett–Packard Graphics Language (HPGL) is quickly becoming a de facto standard with Hewlett–Packard plotters. Most graphics software programs now support plotters that use HPGL.

Products and Companies Mentioned in This Chapter

PRINT SCREEN ($50), DOMUS Software Limited, 251 Cooper St., Ottawa, Ontario K2P 0G2 Canada. (613) 230–6285

PC Paint ($99), Mouse Systems Corp., 2600 San Tomas Expressway, Santa Clara, CA 95051. (408) 988–0211

Graphwriter ($595), Graphics Communications Inc., 200 Fifth Ave., Waltham, MA 02254. (617) 890–8778

Freelance ($395), Graphics Communications Inc., 200 Fifth Ave., Waltham, MA 02254. (617) 890–8778

Chemical Structures for Freelance ($99), BREGO Research, 5989 Vista Loop, San Jose, CA 95124. (408) 723–0947

AutoCAD ($1000–$2500), Autodesk, 2320 Marinship Way, Sausalito, CA 94965. (415) 331–0356

Enhanced Color Display ($849) and Enhanced Graphics Adapter ($524), IBM, 1000 N.W. 51st St., Boca Raton, FL 33432. (800) 447–4700

Hercules Graphics Board ($495), Hercules Computer Technology, 2550 Ninth St., Berkeley, CA 94710. (415) 540–6000

Dazzler 2 ($495), Sigma Designs Inc., 2023 O'Toole Ave., San Jose, CA 95135. (408) 943–9480

AST Research, 2121 Alton Ave., Irvine, CA 92714. (714) 863–1333

Tseng Labs, 205 Pheasant Run, Newtown, PA 18940. (215) 968–0502

Quadram Corp., 4355 International Blvd., Norcross, GA 30093. (404) 923–6666

Plantronics Inc., 1751 McCarthy Blvd., Milpitas, CA 95035. (408) 945–8711

Paradise Systems, 217 E. Grand Ave. South, San Francisco, CA 94080. (415) 588–6000

Chapter

Data Base Management Systems

omputerized *data base management systems* (*DBMSs*) were once only found on mainframe and minicomputers. But as PCs became widely available and more powerful, a large number of powerful programs that manage data on PCs were developed. These programs span a wide spectrum of applications, from individual data file systems, which are as easy to use as a telephone Rolodex, to full-powered relational DBMSs with features equal to those found on mainframes. With this wide range of products, anyone can find a DBMS that fits a needed application like a glove. This means a data base can be generated and used even from floppy disks. An IBM PC/AT with 30 or more megabytes of storage or a local area network (explained more fully in Chapter 9) provides enough disk space and processor speed to operate a full-featured DBMS.

A new generation of DBMSs now on the market takes full advantage of the latest advances in computer hardware and software. Products like Microrim's R:Base 5000, Ashton–Tate's dBase III, and MDBS's Knowledgeman provide mainframe performance at microcomputer prices. Some recent innovations for DBMSs include Microrim's CLOUT, a step toward natural language interfaces; integration of data with graphics in Reflex by Borland/Analytica; integration of data management with spreadsheet data in Lotus 1–2–3; Lotus Symphony; MDBS's Knowledgeman; and joint efforts by Ashton–Tate and Lotus with Informatics to pass data from mainframe data bases into data bases running on the IBM PC. Both R:Base 5000 and dBase III have network-compatible versions of their products. These versions can run on a local area network and can perform the necessary file and record locking for complete data security.

1000–4/87/0155$09.00/1   © 1987 American Chemical Society

Another trend for PC data management is that many mainframe data management programs are now compatible with PCs. These programs, like PC/FOCUS, usually have all the functions of the mainframe program including the same commands; a mainframe user does not have to relearn how to use the program to run it on a PC. This strategy takes advantage of the brainware already learned on the mainframe. This combination of PCs and mainframes bridged by the same software is sure to be popular with many data management users.

What Is a Data Base Management System?

Everyone is familiar with data bases because data bases are used every day. A *data base* is simply a collection of information organized and presented to serve a specific purpose. Put another way, data bases provide data to answer specific questions. Some familiar data bases are the telephone directory, dictionaries, the *Handbook of Chemistry and Physics*, encyclopedias, and the card catalog at the local library. Data bases are unique because information is presented in a relational manner; thus it is easy to obtain the information of interest. In the telephone book, the addresses and telephone numbers are related to a specific name. The names are listed in alphabetical order, and when the desired name is found, the person's address and telephone number are on the same line.

Similarly, in the *Handbook of Chemistry and Physics*, the compounds are listed in alphabetical order by their base name. In the library, the card catalog orders the data by author, subject, and book title.

Computerized data bases can fill data storage and retrieval needs just as simply as the printed ones and have additional capabilities not possible with printed data bases. These major capabilities include quick searching, sorting, and easy reporting of results in the form selected. For example, if only a compound's melting point is known, then tracking down the compound's name or formula using the *Handbook of Chemistry and Physics* would be very difficult because the handbook is organized and sorted by molecular formula and compound name. In a computerized data base, information can be sorted, listed, and searched by using any of the elements in the data base (Figure 7.1).

DBMSs are programs that create the data base and data organization and allow data to be entered, retrieved, and edited. Just creating

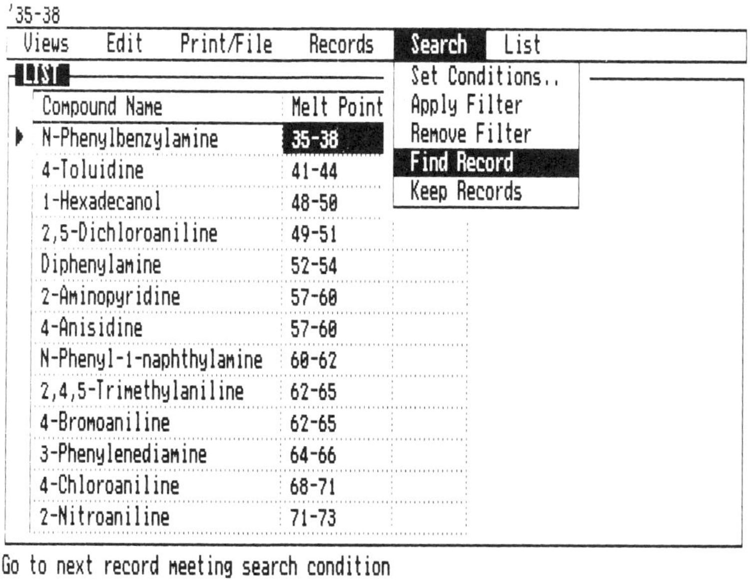

'35-38

| Views | Edit | Print/File | Records | Search | List |

LIST

Compound Name	Melt Point
▶ N-Phenylbenzylamine	35-38
4-Toluidine	41-44
1-Hexadecanol	48-50
2,5-Dichloroaniline	49-51
Diphenylamine	52-54
2-Aminopyridine	57-60
4-Anisidine	57-60
N-Phenyl-1-naphthylamine	60-62
2,4,5-Trimethylaniline	62-65
4-Bromoaniline	62-65
3-Phenylenediamine	64-66
4-Chloroaniline	68-71
2-Nitroaniline	71-73

Search menu:
Set Conditions..
Apply Filter
Remove Filter
Find Record
Keep Records

Go to next record meeting search condition

*Figure 7.1   In a computerized data base, any of the field
values can be searched and sorted.*

a data base and storing the data does little good. All DBMSs allow
users to enter new data, edit the contents of the data base, sort it, and
store it. DBMSs also provide the tools to extract data from the data
base in a meaningful fashion. Two major methods of retrieving data
are report generators for printing standard reports, and ad hoc
queries to obtain specific information whenever necessary.

Report generators, or report writers, allow reports to be created,
usually on the display screen, with the exact information in the
desired location. This report can then be generated with a simple
command. The other type of data-retrieval method is an *ad hoc
query.* Most DBMSs have a query command language that can be
entered to obtain specific information whenever the user so requests.
The syntax used in these queries usually must be exact. In the case of
Microrim's CLOUT, these queries can be more general (less structured), and the user can define new key words to fit the application
more closely.

Data Management Programs for the IBM PC

Data management programs organize data into three basic units: *files,
records,* and *fields.* A common box of index cards is an example of a

noncomputerized data base. Computerized data base terms can be described by using a box of index cards as an analogy. The box itself is the file in our computerized data base. Each card in the index box is a record. Each index card has one or more fields of information written on the card.

Three categories of data base management programs have evolved for use on the IBM PC. These categories differ in the way data is stored and handled, the ease of program use, and the application choices for the retrieved data. The capabilities of the DBMSs must match the desired application. Simple filing applications are the easiest to implement and usually require only a file management program. If data must be integrated with other applications such as word processing, graphics, or spreadsheets, then an integrated package with data management capabilities can be used. Finally, if a large data base application requires many different files of data and possibly a network of computers for data entry and sharing, then a full-featured relational DBMS can be used.

Four data management programs will now be investigated. PC–File II, Reflex, and Lotus 1–2–3 are single-file programs. The fourth program, R:Base 5000, is a relational data base that stores data in many files.

Filing Programs

The simplest type of data management programs are filing programs. *Filing programs* allow data stored in one disk file to be created, edited, and reported. These programs are the electronic equivalent of a card index. As an example, these programs can be used easily to make a data base of chemicals found in a lab. The data base would include compound names, manufacturers, date purchased, some of the compound's physical characteristics, and the amount on the shelf. Data stored in the data base is similar to data stored in a card index. Each chemical has information on a separate card, or record, in the data base. Each piece of information in the record is called a field (Figure 7.2). A data base can be created by using one of many popular data filing programs such as PC–File II.

PC–File II. PC–File II was developed by Jim Button and is distributed by his company, ButtonWare. This program has been widely accepted because of its usefulness and low price. The program is distributed as freeware, which means that it is free to anyone who would like to use

Field Names   Records

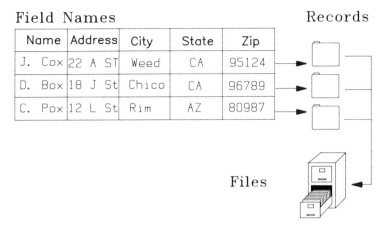

Name	Address	City	State	Zip
J. Cox	22 A ST	Weed	CA	95124
D. Box	18 J St	Chico	CA	96789
C. Pox	12 L St	Rim	AZ	80987

Files

Figure 7.2   Relationship among fields, records, and files.

it. Regular users are asked to make a $25 contribution to support the program further. This method of software distribution has proven very successful for PC–File II. ButtonWare has improved on PC–File II with the introduction of PC–File III and PC–File Relational. PC–File Relational is sold in the usual manner but is still attractively priced, and PC–File III is still being distributed as freeware.

The first step in setting up a chemical storeroom data base is to decide what data to keep. At this point in the data base design, thinking about what kinds of questions the data base will need to answer is important. The data base should help the user keep track of what chemicals are in the storeroom, where they are located, how much is available, and when to reorder a certain chemical. Here is a list of data that will be used in the data base:

Chemical name

Location

Current amount

Manufacturer

Reorder amount

Price

Purchase date

When last used

Last person to use

Creating this data base with PC–File II is done in the following manner: PC–File II is distributed with all the programs needed, and the documentation for the program is on disk in a file named

PC–FILE.DOC. Printing this manual can be done by copying the documentation file to the printer. Place the distribution disk in disk drive A: and enter the command **A>COPY PC-FILE.DOC LPT1:**. As can be seen, the documentation is not long but is complete and easy to read.

Load the program with the command **A>PC-FILE**. Then select the B: disk for data and enter the file name for the data base as **CHEMICAL**. Because the data file is not yet present on the disk, the program asks if the data base entries will be defined. Now the names of the field entries and the number of characters for each can be entered. The original list can be used, and the field names and number of characters can be assigned in the following way:

Chemical name **NAME 30**

Location **LOC 5**

Current amount **CAMT 8**

Manufacturer **VENDOR 10**

Reorder amount **REORDAMT 8**

Price **PRICE 8**

Purchase date **PURDATE 8**

When last used **LASTDAY 8**

Last person to use **LPERSON 10**

Now the data base has been defined. Some data can now be entered. After defining the data base structure, the user is sent to the main menu, where many options can be selected for data entry and reporting. Select to enter data by pressing the F3 key. A data entry form is displayed on the screen, and values can be entered for each field. Enter the following data:

ACETONE,12D,200,Kodak,50,22.5,1/12/86,1/20/86,J. Jones
TOLUENE,11D,300,Kodak,50,34.50,1/13/86,1/18/86,K. Smith
BENZENE,10D,500,Kodak,50,10.50,1/4/86,1/23/86,L. Larson

This data can be printed as shown in Figure 7.3. The data can be sorted by when each chemical was last used (Figure 7.4). This sorting will let the user know which chemicals are used frequently. Actually, another field should be added that tells the user how often a chemical is used. This can be done by adding a new field that will be called NUMUSED (number of times used). Each time someone uses a chemical, this number will be increased by 1. The frequency of chemical use will be known.

If data has been entered into the data base, to add new field data simply use the clone command to output the data. Then define a new data base with the additional fields and read in the old data.

```
06-19-1986 AT 14:50              Chemical Database                              Page 1
name        loc   camt   vendor      reordamt  price   purdate   lastday  lperson
========    ====  =====  ==========  ========  =====   =======   =======  =======
ACETONE     12D   200    Kodak       50        22.5    1/12/86   1/20/86  J. Jones
TOLUENE     11D   300    Kodak       50        34.50   1/13/86   1/18/86  K. Smith
BENZENE     10D   500    Kodak       50        10.50   1/4/86    1/23/86  L. Larson

          TOTALS:
    ----------------------------------------------------------------------------
Printed 3 of the 3 records.
```

Figure 7.3   Printout of the chemical storeroom data.

```
06-19-1986 AT 14:59      Sorted Chemical Data by Last Used Day                  Page 1
name        loc   camt   vendor      reordamt  price   purdate   lastday  lperson
========    ====  =====  ==========  ========  =====   =======   =======  =======
BENZENE     10D   500    Kodak       50        10.50   1/4/86    1/23/86  L. Larson
ACETONE     12D   200    Kodak       50        22.5    1/12/86   1/20/86  J. Jones
TOLUENE     11D   300    Kodak       50        34.50   1/13/86   1/18/86  K. Smith

          TOTALS:
    ----------------------------------------------------------------------------
Printed 3 of the 3 records.
```

Figure 7.4   Data sorted by when each chemical was last used.

When first using a data base, the user will want to add new fields and probably remove some fields. This is a very natural occurrence. For this reason, a simple data base application should be used first; the strategies learned from first experiences will be very valuable in later applications. It is also important to start with easy-to-use programs. The easier the program is to use, the less likely errors will be made because of misunderstanding the requirements of the program.

PC–File II is an excellent program for learning about data management. PC–File II is easy to use, and what it does, it does well.

Reflex. Reflex, which has many functions not available in PC–File II, is distributed by Borland International. Much of Reflex's popularity is due to its ease of use, ability to create different "views" of the stored data, and very low price ($95). The program also comes with a full-function report writer and many of the features found in higher cost products. In fact, until recently Reflex sold for $495.

CREATING THE DATA BASE. We will create a chemical data base, but this time physical properties of the chemicals will also be stored. This data base will be used to find chemicals that have a specific boiling point or refractive index. Setting up the chemical data base is easy. First, write out the desired traits of each compound, including the number of characters reserved for each property, and the type of value (number, integer, or text) to store (Figure 7.5). Now start the program by entering the command **A>REFLEX**. The main program menu should now be presented (Figure 7.6).

Select to create a new data base. Then enter the information about the number, size, and type of each field in the data base. Figure 7.7 shows the data base setup using the form view. The same data can also be viewed in the list view (Figure 7.8) and edited.

ENTERING DATA. The data can be entered manually or loaded into the data base from disk files with the proper formats (Figure 7.9). Enter data by filling in the form displayed on the screen. Advanced data management programs such as Reflex have the ability to re-create paper forms on the computer screen. The form view of the data allows labels and data entry points to be placed anywhere on the screen.

EDITING DATA. Data can be reviewed or edited using the edit function on the same forms just created. More data records can be

'Acetone

| Views | Edit | Print/File | Records | Search | Form |

FORM

Solvent Database

Name: Acetone

Location: R10

Boiling Point deg C: 56   Dielectric Constant: 20.70

Density 25 deg C: 0.786   Refractive Index: 1.3590

Flash Point C: -18   Flash Point F: -0

Price 4x4L: $23.90   Price Spec Grade: $25.50

Figure 7.5   Data to be stored in the physical properties data base.

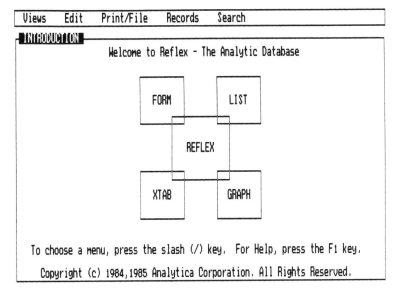

Figure 7.6   Reflex main menu. Selection of functions is made easy by moving over menus with the cursor keys or by using a mouse.

```
 Views    Edit    Print/File    Records    Search    Form
FORM DESIGN
                          'Solvent Database
 Name
 Location
 Boiling Point deg C              Dielectric Constant
 Density 25 deg C  ▮              Refractive Index
 Flash Point C                    Flash Point F
 Price 4x4L                       Price Spec Grade



 Line: 008 Col: 019
```

Figure 7.7   Setting up the data base in Reflex by using the form screen.

```
1.359
 Views    Edit    Print/File    Records    Search    List
LIST
   Name         Locat  Boil  Densit  Diele  Refract  Flas  Price 4x  Price Sp
 ▶ Acetone      R10      56   0.786   20.70  1.3590    -18   $23.90    $25.50
   Chloroform   R4       61   1.471    4.81  1.4475           $48.50    $49.94
   Carbon Tet   R5       77   1.583    2.24  1.4631           $43.50    $51.95
   Cyclohexan   R6       81   0.772    2.02  1.4263    -20   $31.55    $41.80
   Acetonitri   R14      82   0.775   37.50  1.3441      8   $39.60    $68.50
   Butyl Alco   R1      118   0.806   17.51  1.3993     29   $31.50
   Acetic Anh   R12     140   1.069   20.70  1.3904     53   $54.50
   Aniline      R3      184   1.018    1.59  1.5863     76   $67.80
```

Figure 7.8   Setting up the data base using the list view.

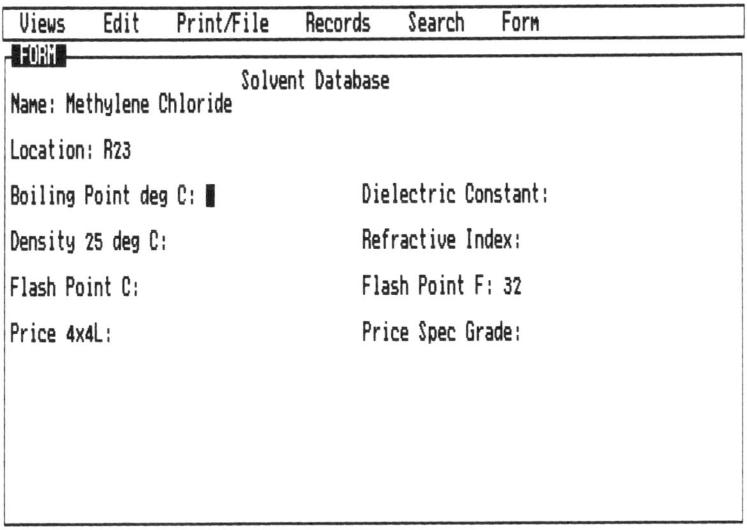

```
┌──────────────────────────────────────────────────────────┐
│  Views    Edit    Print/File    Records    Search    Form  │
├──────────────────────────────────────────────────────────┤
│ ▄FORM▄───────────────────────────────────────────────────│
│                         Solvent Database                   │
│ Name: Methylene Chloride                                   │
│                                                            │
│ Location: R23                                              │
│                                                            │
│ Boiling Point deg C: █          Dielectric Constant:       │
│                                                            │
│ Density 25 deg C:               Refractive Index:          │
│                                                            │
│ Flash Point C:                  Flash Point F: 32          │
│                                                            │
│ Price 4x4L:                     Price Spec Grade:          │
│                                                            │
│                                                            │
│                                                            │
│                                                            │
└──────────────────────────────────────────────────────────┘
```

Figure 7.9   Entering data using the form view. Paper forms can be re-created on the screen.

added to the data base by filling in the blanks on the developed forms.

SEARCHING AND SORTING DATA.   Data can also be searched and sorted. These applications of computerized data bases compensate for all the time users spend organizing information. Suppose you are working on the synthesis of compound TX45Z, and you need a solvent that boils in the 65–75 °C range. By listing the solvents in the data base ordered by boiling point, you can easily pick a good candidate for the desired solvent.

TAKING DIFFERENT VIEWS OF THE DATA.   One of Reflex's major attractions is ability to produce many views of the data in the data base. These views include List, Graph, Form, and Crosstab. Each of these ways of looking at data has broad applications. Figures 7.10–7.13 show different views of current data. Views can be mixed and matched; as many as three different views of a file can be displayed on the screen at one time.

REFLEX REPORT WRITER.   Reflex has an excellent report writer. With the report writer, almost any type of printed report can be generated.

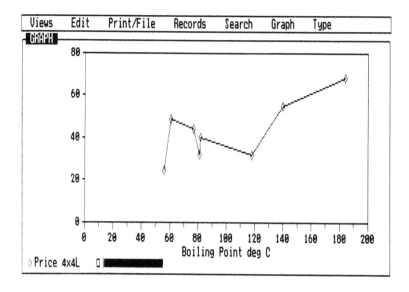

Figure 7.10   Graph view of the data. Many types of graphs can be generated, including bar, multiple bar, line, pie, and scatter charts.

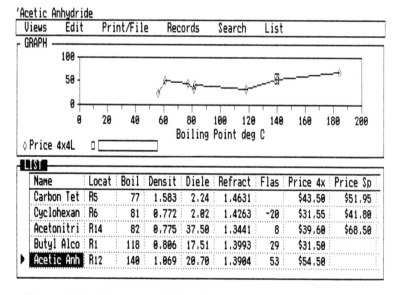

Figure 7.11   List view of the data. Lists can be made in any order and format.

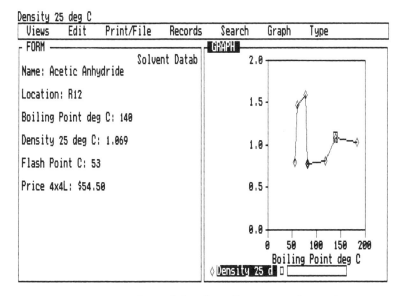

Figure 7.12   Form view of the data. Forms are the easiest way to enter data.

Name = 'Acetic Anhydride'

Views	Edit	Print/File	Records	Search	Crosstab

CROSSTAB

Summary: [@AVG]   Field: [Boiling Point deg C]

Flash Point C

		-20	-18	8	29	53	76	ALL
N	Acetic					140		140
a	Acetone		56					56
m	Acetoni			82				82
e	Aniline						184	184
	Butyl A				118			118
	Carbon							77
	Chlorof							61
	Cyclohe	81						81
	ALL	81	56	82	118	140	184	100

Figure 7.13   Crosstab view of the data. This view allows interrelationships to be found easily among the pieces of data.

The report writer can also be used to create disk files of the data with specific formats. This report writer is very useful for sharing data with other programs because Reflex will format the file exactly as needed. The report writer utilities can also be used to define incoming data so data from other programs can be loaded correctly into the data base.

Lotus 1–2–3: Data Management within Integrated Applications. Most of the popular integrated application programs include data management as one of their functions. The data management features included in these programs are normally at the same level as filing programs. A data base can be created, and data can be entered, edited, searched, and sorted. Most integrated programs include a report writer for generating custom reports. The real value of these systems is what can be done with the data once it has been entered. Most DBMS functions stop once a report or results of a query have been printed or displayed. The integrated software functions start at this point.

Here is an example of using the data management commands found in Lotus 1–2–3 in conjuction with spreadsheet calculations and graphics. This example sets up a complete sample tracking system and uses data management commands to extract data from the main data base, sort the data, and generate a graph of the results.

DATA ANALYSIS LOGBOOK. The *logbook* can be used to track the results of tests performed on final or intermediate manufactured products. The logbook can be modified easily to perform a number of functions. Data is entered into the logbook in two ways, manually through the keyboard and by importing data files captured directly from the instruments. This data can then be analyzed, correlated, and summarized by using Lotus 1–2–3's data management and graphics commands.

After some data has been captured, the logbook can be set up for the results of the analysis. The logbook has two parts, data log and data query. The *data log* is a list of the sample analysis information, including a reference number, date the sample was received, who submitted the sample, type of analysis, analyte or feature measured in the sample, analysis raw value, computed result, date analyzed, and a memo for other information. The *data query* section is a method used for summarizing and extracting data from the data log information (data base).

The following example for setting up the logbook uses the same terminology as the Lotus 1–2–3 instruction manual for the various commands. This example was first published in *PCWorld* in February 1984.

The first step is to build the data log input form. Make sure a clean worksheet is used by giving the command /**Worksheet Erase Yes**.

In cell A1 enter *** **Laboratory Analysis Logbook** ***. Move to A5 and enter **Data Log**.

Place a repeating label of equal signs in A6 by entering \=. Copy the contents of A6 to cells B6–I6 with the command **CopyA6** <**ENTER**>**B6..I6**<**ENTER**>.

Now enter the field names in row 7: in A7 enter "**Ref no.**, in B7 enter "**Rec Date**, in C7 enter **Submitted by**, in D7 enter **Analysis**, in E7 enter **Analyte**, in F7 enter "**Raw Value**, in G7 enter **Amount**, in H7 enter **Date Anal**, and in I7 enter **Memo**.

To complete these entries, go to A8 and make a repeating line of hyphens by entering \-. Use the copy command to complete the line from B8 through I8.

Now reset the column width of column C to 20 characters. Go to any cell in column C and give the command /**Worksheet Column Set 15** <**ENTER**>.

Using the same command, set the column widths of columns D and E to 8 characters each. Set columns F and G to 12 characters, column H to 9 characters, and column I to 15 characters.

Most of the numbers on the worksheet will be displayed to two decimal places. Set the entire worksheet for this format with the command /**Worksheet Global Format Fixed2** <**ENTER**>.

The "**Ref no.** and "**Raw Value** columns should display their data as integers. Reformat columns A and F as fixed with no decimal places with the command /**Range Format Fixed0** <**ENTER**>**A9..A500** <**ENTER**>. The worksheet should look like Figure 7.14. Save the work with the command /**File Save PCLAB**<**ENTER**>.

We can enter some data from a typical chemical analysis laboratory to show how the logbook can be used. This lab performs six types of analyses: gas chromatography (**GCHROM**), liquid chromatography (**LCHROM**), inductively coupled plasma spectroscopy (**ICP**), ultraviolet spectroscopy (**UV**), infrared spectroscopy (**IR**), and gel permeation chromatography (**GPC**).

These six analyses are performed on seven compounds or elements: propane, aspirin, mercury (Hg), butane, polyethylene (vinyl),

```
       A         B          C          D    E      F        G        H          I
1   *** Laboratory Analysis Logbook ***
2
3
4
5   Data Log
6   ============================================================================================
7   Ref no. Rec Date Submitted by      Analysis Analyte Raw Value   Amount    Date Anal     Memo
8
```

Figure 7.14   Worksheet ready for data entry.

methane, and ethane. When a sample is received, a reference number, date received, the submitter's name, analysis type, and analyte are entered into the data log. When the analysis is completed, the raw value, amount, and date analyzed are filled in along with any comments in the memo section. When calibration runs are performed on an instrument, the results are also logged. The submitter's name is replaced with **Calibration** for these entries.

All dates are entered with the first two digits representing the year, the next two digits the month, and the decimal portion the day. This method of representing the date is very practical when using the 1–2–3 data query and sort commands because all dates stay in numerical order. The only negative feature is that this method of representation cannot easily compute the number of days between two dates.

Enter the data as shown in Figure 7.15. At this point the worksheet should be resaved with the command /**File Save PCLAB<ENTER> Replace.**

The next steps will introduce some unique features of 1–2–3. Using the data log with the 1–2–3 data query and sort commands requires names to be assigned to various ranges of cells on the worksheet. First, the entire data log range is given the name INPUT by going to cell A7 and giving the command /**Range Name Create INPUT <ENTER>A7..I500<ENTER>.**

This command names the range of cells from A7 to I500 as INPUT and sets the lower boundary of the data log at row 500. Note the range started on row 7 to include the labels given at the top of each field. These labels must be included in the input range for the data query and sort commands to work correctly.

Now the data query section can be set up. Go to cell J1 and enter "QUERY. Then go to J2 and place a repeating label of equal signs by entering \=. Copy the contents of J2 to K2–R2 with the command /**CopyJ2<ENTER>K2..R2<ENTER>.**

	A	B	C	D	E	F	G	H	I
1	*** Laboratory Analysis Logbook ***								
2									
3									
4									
5	Data Log								
6	==								
7	Ref no.	Rec Date	Submitted by	Analysis	Analyte	Raw Value	Amount	Date Anal	Memo
8	101	8308.01	Smith	GCHROM	propane	201045	400.00	8308.04	
9	102	8308.01	Jones	LCHROM	asprin	45346	385.00	8308.04	
10	103	8308.02	Green	ICP	Hg	2314560	25.60	8308.05	
11	104	8308.03	Brown	ICP	Hg	2150780	25.50	8308.05	
12	105	8308.03	Johnson	UV	asprin	35245	340.00	8308.07	
13	106	8308.04	Edger	IR	asprin	40563	358.00	8308.07	
14	107	8308.04	Johnson	GCHROM	butane	23568	229.86	8308.08	
15	108	8308.04	Weaver	GPC	polyv	342789	469.23	8308.08	
16	109	8308.05	Mills	ICP	Hg	1894670	25.30	8308.08	
17	110	8309.06	Peterson	GPC	vinyl	3456790	456.00	8308.10	New Entry
18	2001	8307.05	Calibration	GCHROM	propane	230145	500.00	8307.05	
19	2002	8307.05	Calibration	GCHROM	methane	150032	500.00	8307.05	
20	2003	8307.05	Calibration	GCHROM	ethane	120132	500.00	8307.05	
21	2004	8307.12	Calibration	GCHROM	propane	234024	500.00	8307.12	
22	2005	8307.12	Calibration	GCHROM	methane	160054	500.00	8307.12	
23	2006	8307.12	Calibration	GCHROM	ethane	124098	500.00	8307.12	
24	2007	8307.19	Calibration	GCHROM	propane	228023	500.00	8307.19	
25	2008	8307.19	Calibration	GCHROM	methane	145045	500.00	8307.19	
26	2009	8307.19	Calibration	GCHROM	ethane	118043	500.00	8307.19	
27	2010	8307.26	Calibration	GCHROM	propane	232076	500.00	8307.26	
28	2011	8307.26	Calibration	GCHROM	methane	155076	500.00	8307.26	
29	2012	8307.26	Calibration	GCHROM	ethane	112043	500.00	8307.26	
30	2013	8308.02	Calibration	GCHROM	propane	227032	500.00	8308.02	
31	2014	8308.02	Calibration	GCHROM	methane	130102	500.00	8308.02	
32	2015	8308.02	Calibration	GCHROM	ethane	107021	500.00	8308.02	
33	2016	8308.09	Calibration	GCHROM	propane	225021	500.00	8308.09	
34	2017	8308.09	Calibration	GCHROM	methane	120034	500.00	8308.09	
35	2018	8308.09	Calibration	GCHROM	ethane	100103	500.00	8308.09	
36	2019	8308.16	Calibration	GCHROM	propane	220012	500.00	8308.16	
37	2020	8308.16	Calibration	GCHROM	methane	125013	500.00	8308.16	
38	2021	8308.16	Calibration	GCHROM	ethane	97035	500.00	8308.16	
39	2022	8308.23	Calibration	GCHROM	propane	215078	500.00	8308.23	
40	2023	8308.23	Calibration	GCHROM	methane	110021	500.00	8308.23	
41	2024	8308.23	Calibration	GCHROM	ethane	94002	500.00	8308.23	
42	2025	8308.30	Calibration	GCHROM	propane	210037	500.00	8308.30	
43	2026	8308.30	Calibration	GCHROM	methane	90103	500.00	8308.30	
44	2027	8308.30	Calibration	GCHROM	ethane	90106	500.00	8308.30	
45	2028	8309.06	Calibration	GCHROM	propane	208016	500.00	8309.06	
46	2029	8309.06	Calibration	GCHROM	methane	74091	500.00	8309.06	
47	2030	8309.06	Calibration	GCHROM	ethane	82001	500.00	8309.06	

Figure 7.15   Data for the data base.

Copy the labels from row 7 of the data log to cells J3–R3 with the command /**CopyA7..I7<ENTER>J3..R3<ENTER>**.

Now make a copy of cells J2–R3 in cells J6–R7 with the command /**CopyJ2..R3<ENTER>J6..R7<ENTER>**.

Reset the column widths to match those in the data log section. The query section should look like Figure 7.16.

The data query and sort commands can be used after a few more named ranges are created. First, set up the criterion, or selection range, by naming cells J3–R4 CRITERION with the command /**Range Name CreateCRITERION<ENTER>J3..R4<ENTER>**.

Then name cells J7–R500 OUTPUT with the command /**Range Name CreateOUTPUT<ENTER>J7..R500<ENTER>**.

Finally, name cells J8–R500 SORT with the command /**Range Name CreateSORT<ENTER>J8..R500<ENTER>**.

The criterion and output ranges include a row of labels, but the sort range does not.

Now all the named ranges can be connected together so they can be used with 1–2–3's data query and sort commands. An input, an output, a sort, and a criterion range must now be defined by first giving the command /**Data Query InputINPUT<ENTER>**, which assigns the A7–I500 range as the input range. Continue in the data submenu with the command **CriterionCRITERION<ENTER>**, which assigns the range J3–R4 as the criterion range. The command **OutputOUTPUT<ENTER>** assigns the range J7–R500 as the output range. Now press the Esc key to get back to the main data menu. Finally, the command **Sort Data-RangeSORT<ENTER>** assigns the range J8–R500 as the sort range.

The data query command allows the user two ways to select items from the data log. The simplest is the find method. Suppose each of the results needs to be viewed by using ICP analysis. Go to cell M4 and enter **ICP** under the analysis label. Now the command /**Data**

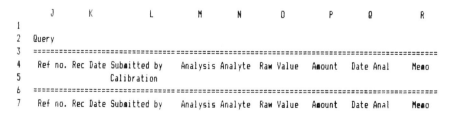

	J	K	L	M	N	O	P	Q	R
1									
2	Query								
3	===
4	Ref no.	Rec Date	Submitted by	Analysis	Analyte	Raw Value	Amount	Date Anal	Memo
5			Calibration						
6	===
7	Ref no.	Rec Date	Submitted by	Analysis	Analyte	Raw Value	Amount	Date Anal	Memo

Figure 7.16 Query section of the worksheet.

Query Find will switch the screen to the data log area (input range), and the third entry will be highlighted. This is the first record in the input range that has a match with the ICP criterion. Now press the cursor down key. The highlight bar jumps to the next match, which is Green's ICP submitted sample. Pressing the cursor down key again jumps the bar to Mills's ICP sample, and pressing the cursor up key jumps the bar back to Green's ICP sample. After looking at the entries, press the Esc key and the screen will return to the query menu.

The other method of data query is extraction. Matching data is displayed in the data output range. As an example of this function, the ICP analysis data will be extracted. Enter the command / **Data Query Extract**.

The matching data entries are displayed as shown in Figure 7.17. All of the calibration data can now be extracted. Go to M4 and erase this cell with the command / **Range Erase<ENTER>**.

Go to L4 and enter **Calibration**. Then enter the command / **Data Query Extract**. In a few seconds all of the entries with **Calibration** will appear in the output range. Values could have been entered in other labeled fields, such as methane in the analyte column, and only those entries with both **Calibration** and **methane** would be extracted. Numeric values and formulas can also be used for the match criterion. For example, entering the formula +**F87>2000000** would yield only three entries (each of the ICP runs) because these entries are the only ones with raw values greater than 2 million.

Experiment with the query capabilities of 1–2–3. Note that once the matching criterion is set up, the query key, which is the F7 key, can be pressed to obtain the queried data.

	J	K	L	M	N	O	P	Q	R
1									
2	Query								
3	========								========
4	Ref no.	Rec Date	Submitted by	Analysis	Analyte	Raw Value	Amount	Date Anal	Memo
5				ICP					
6	========								========
7	Ref no.	Rec Date	Submitted by	Analysis	Analyte	Raw Value	Amount	Date Anal	Memo
8	103	8308.02	Green	ICP	Hg	2314560	25.60	8308.05	
9	104	8308.03	Brown	ICP	Hg	2150780	25.50	8308.05	
10	109	8308.05	Mills	ICP	Hg	1894670	25.30	8308.08	
11									

Figure 7.17 Matched data is displayed in the output range on the spreadsheet.

Extracted data can also be sorted and graphed. With the calibration data in the output range, go to cell N8 and enter /**Data Sort Primary-Key Descending**<ENTER>. This command assigns the analyte column as the primary key.

Now go to cell K8 and type the command **Secondary-Key Ascending**<ENTER>. This command assigns the received date as the secondary key. Because the sort range has already been defined, just enter **Go**. This procedure sorts and graphs the extracted data.

Before the sort, the calibration data is ordered as shown in Figure 7.18. The entire list of 30 entries will be sorted and grouped by analyte, with each set of analytes ordered by the date analyzed, as shown in Figure 7.19.

	J	K	L	M	N	O	P	Q	R
1									
2	Query								
3	==								
4	Ref no.	Rec Date	Submitted by	Analysis	Analyte	Raw Value	Amount	Date Anal	Memo
5			Calibration						
6	==								
7	Ref no.	Rec Date	Submitted by	Analysis	Analyte	Raw Value	Amount	Date Anal	Memo
8	2001	8307.05	Calibration	GCHROM	propane	230145	500.00	8307.05	
9	2002	8307.05	Calibration	GCHROM	methane	150032	500.00	8307.05	
10	2003	8307.05	Calibration	GCHROM	ethane	120132	500.00	8307.05	
11	2004	8307.12	Calibration	GCHROM	propane	234024	500.00	8307.12	
12	2005	8307.12	Calibration	GCHROM	methane	160054	500.00	8307.12	
13	2006	8307.12	Calibration	GCHROM	ethane	124098	500.00	8307.12	
14	2007	8307.19	Calibration	GCHROM	propane	228023	500.00	8307.19	
15	2008	8307.19	Calibration	GCHROM	methane	145045	500.00	8307.19	
16	2009	8307.19	Calibration	GCHROM	ethane	118043	500.00	8307.19	
17	2010	8307.26	Calibration	GCHROM	propane	232076	500.00	8307.26	
18	2011	8307.26	Calibration	GCHROM	methane	155076	500.00	8307.26	
19	2012	8307.26	Calibration	GCHROM	ethane	112043	500.00	8307.26	
20	2013	8308.02	Calibration	GCHROM	propane	227032	500.00	8308.02	
21	2014	8308.02	Calibration	GCHROM	methane	130102	500.00	8308.02	
22	2015	8308.02	Calibration	GCHROM	ethane	107021	500.00	8308.02	
23	2016	8308.09	Calibration	GCHROM	propane	225021	500.00	8308.09	
24	2017	8308.09	Calibration	GCHROM	methane	120034	500.00	8308.09	
25	2018	8308.09	Calibration	GCHROM	ethane	100103	500.00	8308.09	
26	2019	8308.16	Calibration	GCHROM	propane	220012	500.00	8308.16	
27	2020	8308.16	Calibration	GCHROM	methane	125013	500.00	8308.16	
28	2021	8308.16	Calibration	GCHROM	ethane	97035	500.00	8308.16	
29	2022	8308.23	Calibration	GCHROM	propane	215078	500.00	8308.23	
30	2023	8308.23	Calibration	GCHROM	methane	110021	500.00	8308.23	
31	2024	8308.23	Calibration	GCHROM	ethane	94002	500.00	8308.23	
32	2025	8308.30	Calibration	GCHROM	propane	210037	500.00	8308.30	
33	2026	8308.30	Calibration	GCHROM	methane	90103	500.00	8308.30	
34	2027	8308.30	Calibration	GCHROM	ethane	90106	500.00	8308.30	
35	2028	8309.06	Calibration	GCHROM	propane	208016	500.00	8309.06	
36	2029	8309.06	Calibration	GCHROM	methane	74091	500.00	8309.06	
37	2030	8309.06	Calibration	GCHROM	ethane	82001	500.00	8309.06	

Figure 7.18   Data before it is sorted.

Now this grouped data can be graphed. Escape back to the main menu by pressing the Esc key. Select the data to plot with /**Graphics A-Data**<**ENTER**>**O8..O17**<**ENTER**>.

This command selects the A range for the propane data. Similarly, the B and C ranges are selected for methane and ethane with the commands **B-Data**<**ENTER**>**O18..O27**<**ENTER**> and **C-Data** <**ENTER**>**O28..O37**<**ENTER**>, respectively.

Now a bar graph with color can be selected and the graph viewed with the command **Type Bar-Graph**<**ENTER**>**Options Color** <**ENTER**>**View**.

By adding legends and titles, a graph like that shown in Figure 7.20 is created.

	J	K	L	M	N	O	P	Q	R
1									
2	Query								
3	==								
4	Ref no.	Rec Date	Submitted by	Analysis	Analyte	Raw Value	Amount	Date Anal	Memo
5			Calibration						
6	==								
7	Ref no.	Rec Date	Submitted by	Analysis	Analyte	Raw Value	Amount	Date Anal	Memo
8	2001	8307.05	Calibration	GCHROM	propane	230145	500.00	8307.05	
9	2004	8307.12	Calibration	GCHROM	propane	234024	500.00	8307.12	
10	2007	8307.19	Calibration	GCHROM	propane	228023	500.00	8307.19	
11	2010	8307.26	Calibration	GCHROM	propane	232076	500.00	8307.26	
12	2013	8308.02	Calibration	GCHROM	propane	227032	500.00	8308.02	
13	2016	8308.09	Calibration	GCHROM	propane	225021	500.00	8308.09	
14	2019	8308.16	Calibration	GCHROM	propane	220012	500.00	8308.16	
15	2022	8308.23	Calibration	GCHROM	propane	215078	500.00	8308.23	
16	2025	8308.30	Calibration	GCHROM	propane	210037	500.00	8308.30	
17	2028	8309.06	Calibration	GCHROM	propane	208016	500.00	8309.06	
18	2002	8307.05	Calibration	GCHROM	methane	150032	500.00	8307.05	
19	2005	8307.12	Calibration	GCHROM	methane	160054	500.00	8307.12	
20	2008	8307.19	Calibration	GCHROM	methane	145045	500.00	8307.19	
21	2011	8307.26	Calibration	GCHROM	methane	155076	500.00	8307.26	
22	2014	8308.02	Calibration	GCHROM	methane	130102	500.00	8308.02	
23	2017	8308.09	Calibration	GCHROM	methane	120034	500.00	8308.09	
24	2020	8308.16	Calibration	GCHROM	methane	125013	500.00	8308.16	
25	2023	8308.23	Calibration	GCHROM	methane	110021	500.00	8308.23	
26	2026	8308.30	Calibration	GCHROM	methane	90103	500.00	8308.30	
27	2029	8309.06	Calibration	GCHROM	methane	74091	500.00	8309.06	
28	2003	8307.05	Calibration	GCHROM	ethane	120132	500.00	8307.05	
29	2006	8307.12	Calibration	GCHROM	ethane	124098	500.00	8307.12	
30	2009	8307.19	Calibration	GCHROM	ethane	118043	500.00	8307.19	
31	2012	8307.26	Calibration	GCHROM	ethane	112043	500.00	8307.26	
32	2015	8308.02	Calibration	GCHROM	ethane	107021	500.00	8308.02	
33	2018	8308.09	Calibration	GCHROM	ethane	100103	500.00	8308.09	
34	2021	8308.16	Calibration	GCHROM	ethane	97035	500.00	8308.16	
35	2024	8308.23	Calibration	GCHROM	ethane	94002	500.00	8308.23	
36	2027	8308.30	Calibration	GCHROM	ethane	90106	500.00	8308.30	
37	2030	8309.06	Calibration	GCHROM	ethane	82001	500.00	8309.06	

Figure 7.19   Data after the sort.

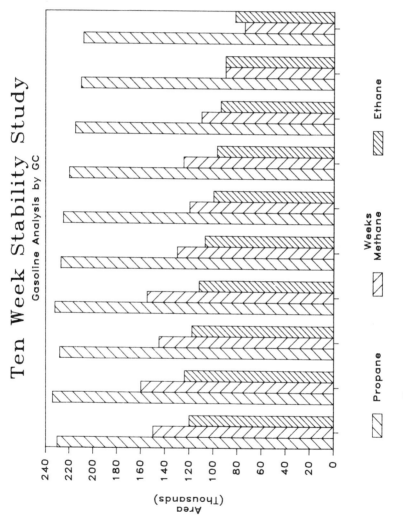

Figure 7.20   Graph of the calibration data in Figures 7.18 and 7.19.

NUMEROUS APPLICATIONS. Once 1-2-3 has been used for a few data management and graphics applications, many more applications can be thought of for this versatile program. A number of major features of the program were not mentioned. For example, tables of data can be generated using the data tables command (see page 205 of the 1-2-3 instruction manual, or *PC World*, Vol. 1, number 6, page 192), or keystroke macros can be used to program a worksheet (see page 107 of the 1-2-3 manual). Handling data in 1-2-3 is perfect for first-time interfacing with instruments because the design of the data reduction and display can be performed in the program. If special data reduction or display is not available in 1-2-3, a custom program can be written.

Lotus has also made available the 1-2-3 Report Writer, which is one of many Lotus add-on programs. Report Writer allows reports to be created from the Lotus data by using techniques usually found only in more advanced DBMSs.

Heavyweight Data Base Applications

Filing programs and the data management tools in integrated programs can solve a large proportion of computerized data management needs. For many, these tools are the only ones needed. A third group of data management programs is available on PCs that are normally referred to as heavyweights. These programs have been patterned after the data base systems found on mainframe computers and can perform like them. In many cases, the data management programs found on PCs are more powerful than their mainframe ancestors because the latest innovations in software are developed and practiced on PCs.

These heavyweight programs have all the functions of simpler filing programs. These programs also have more advanced report generators and more advanced methods of entering data, and most have a procedural language, so custom applications can be programmed. These procedural languages, like regular programming languages, can be programmed to make decisions and automate data processing. Unlike regular programming languages, procedural languages have commands that can retrieve or store data in data bases. Many heavyweight products, such as dBase III and R:Base 5000, have program generators, so even a novice can write programs by making menu selections and filling in the blanks.

Heavyweight applications involve more data, people, and procedures than do filing or integrated programs. Heavyweight data bases are used to serve laboratories of 50 scientists or small companies. More planning is required before the first data base is created, and these applications need more powerful data base tools. Solving large data management needs by conceptually breaking them into many smaller needs or modules is easier than solving the whole problem at once. Heavyweight programs provide this strategy with a relational data base.

Relational DBMSs have distinct advantages over less powerful filing systems. Rather than one data base with a fixed number of fields and all data stored in a single file, a relational data base allows many data bases and files to be related by forming relationships between a field in one data base to one or more fields in other data bases. When new data fields must be added to a relational data base, simply create another data base and form a relationship between the new data base and the existing ones. Adding a new field to a file management system requires unloading of data, changing the form of the data base, and then reloading the old data into the new data base.

Designing a Data Base

Designing a useful heavyweight data base first requires a data model of the desired application to be built. Because three data bases were built earlier in this chapter, many of the mechanics of data management have already been performed. Now the design and planning aspects of data management will be considered in detail.

Preparing a Data Model

Physical models are made all the time to test theories on design before the real thing is made. The data base model will be approached in the same way. A working paper model will yield the blueprints for a heavyweight data base.

The following example can be used for the model. Suppose a product-testing laboratory has testing stations, technicians, and products to test from various departments in the company. Almost any question about the operation of the testing lab will involve these people and objects. Tables of information about these people and objects can be created easily. Figure 7.21 shows a simple data model for the laboratory.

Employee Information

First Name	Last Name	Work in the Lab	Employee No.
Joe	Johnson	12/4/82	5
Pete	Peterson	11/18/81	3
Heidi	Hill	3/29/83	23

Test Station Identification

Test Station No.	Name	Last Maintenance
1	Strength	5/2/84
2	Buoyancy	4/6/84
3	Resistance	3/5/84
4	Strength	5/3/84

Test Data

Test Station No.	Log	Department	Employee No.	Widget Type	Start	End
2	803	R & D	5	Red	2/4/84	3/4/84
3	804	Manufact	23	Blue	3/2/84	3/28/84
1	805	R & D	23	Green	3/4/84	4/4/84
4	806	Manufact	3	Red	2/4/84	3/4/84

Figure 7.21 Data organized as tables. Three tables are defined. The first table contains information about each employee. More fields, or columns, of data could be added easily, such as home address and telephone numbers. The second table contains data to identify each test station. The employee and test station data will not change often. The last table contains the bulk of the test data. The employee number and product test station number are present in the test data. Through this relationship, the data from the other tables can be related to the product test data.

Even with this simple data model of the lab, a number of questions can be answered, such as the following:

- How long does it take to run all the tests on a product?
- What was the average time spent to test a product in March 1984?
- Which technician has performed the most tests?
- Which technician has been in the laboratory the longest?
- When was the last preventive maintenance performed on the equipment?
- What department submits the most products to be tested?

- What is the percent standard deviation of all results at test station 3 with employee 5?

Notice that all of the data components in the model are shown as rows and columns of information. Every row has the same number of entries, one for each column. In each column, the same type of data is collected for each row entry. The actual order of the columns or rows is not important.

In DBMS terminology, each row in this example is called a record, or *tuple*, each column is called a field, or *attribute*, and the entire table is called a file, or *relation*.

Steps in Data Base Design

The three basic steps in designing a data base are listing the objects or people in the working environment, deciding what facts are important, and representing the relationships among the people or objects.

Listing the objects or people in the working environment has been done. There are three tables of objects or people: technicians, testing stations, and products.

To decide what facts are useful or important, a lot of thought is required about what questions need to be answered with the data in the data base. If certain information is not expected to be needed, it does not belong in the data base; if more information for the model will be needed, it should be added to the data base.

Representing the relationships among objects or people requires defining the relationship between these objects or people as one-to-one, one-to-many, or many-to-many. The relationship between technicians and testing stations is one-to-many because one technician can perform at many testing stations. The relationship between the technicians and products is one-to-many because one technician can test many products. Products and testing stations have a many-to-many relationship because many products can be tested at many test stations. A diagram of these relationships is shown in Figure 7.22.

As in other data bases, names must be chosen to represent each of the field values and to identify which kind of data (integer, text, or real) they are. The field names and types of data for this application are shown in Table 7.1.

With this design information, creating a data base for the application described is possible by using one of the popular heavyweight DBMSs. Even if an error is made in designing the data base, data or relationships between data can be added or removed easily.

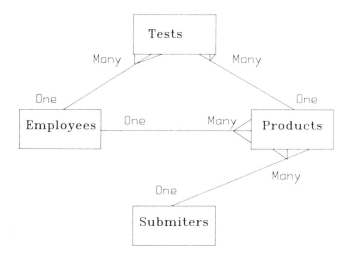

Figure 7.22 Diagram of relationships among data base entries.

Table 7.1   Data Base Field Names, Attributes, and Assigned Tables

Field (column)	Data	Characters	Table
EmpID	Text	5	Roster
LastName	Text	15	Roster
FrstName	Text	15	Roster
HireDate	Date		Roster
StatID	Text	5	TStation
StatType	Text	15	TStation
LastMain	Date		TStation
Product	Text	15	ProdLog
EmpID	Text	5	ProdLog
StrtDate	Date		ProdLog
EndDate	Date		ProdLog
Submiter	Text	15	ProdLog
Result	Real		ProdLog
StatNum	Real		ProdLog
LogNum	Text	5	ProdLog

Implementation Using R:Base 5000

R:Base 5000 has a unique implementation aid called application express. Once the design information is together, implementation is easy. When application express is entered, a screen is presented that contains blanks to be filled in with field names and attributes. The program walks the user through the steps necessary to set up the data base by prompting the user for the necessary information.

Defining the Data Base Structure. Load R:Base, then select application express from the main menu. When the express main menu appears, select item 1, which is defining a new data base. When prompted for the data base name, enter **TESTLAB**. A table definition screen will be displayed. Enter **Roster** for the table name, then **EmpID** for the first column name. When you press the Enter key, application express will display six data types (**Text, Integer, Real, Dollar, Date,** and **Time**). Choose **Text** for the data type for this column. **Integer** could have also been selected, but because calculations will not be performed with this value, it is better to use **Text**. Text can be stored in fewer bytes, and the user will have more flexibility in searching for matches because wild card characters can be used for searching text strings. (A wild card character allows a match with any other characters. The asterisk [*] is usually used as a wild card character for one or more characters, and the question mark [?] is used for one character. For example, if you wish to find all the names in a data base that start with the letter *G*, simply query for names with **G***. This command will show all people with names starting with *G*.) Now the rest of the **Roster** table data can be entered. Press the Esc key when the last definition has been entered.

Continue creating the other two tables, TStation and ProdLog. When a column name is entered that was defined in a previous table, application express automatically fills in the data type and length. If any errors are made, the F2 key can be used to delete a column and the F1 key used to insert a column. When finished defining the tables, exit to the application express main menu.

Building Application Menus. Application express also allows menus to be set up so people not interested in the underlying mechanics of the program can enter data, retrieve data, and print reports. When the implementation session with application express is complete, the program creates another program written in the R:Base

5000 procedural language. This program can then be used as written, or if necessary modified or customized to perform other tasks. Using the procedural language, very complex applications can be created.

An application created using application express consists of menus and submenus with as many as three levels. Menus can be horizontal or vertical. Vertical is best for menus with nine items or fewer. Figure 7.23 shows the menus designed for this application. Each menu item either selects another menu or performs a specific R:Base command. R:Base can perform six basic actions: load, edit, delete, browse, select, or print.

For loading data, application express will lead the user through the design of a simple vertical form for data entry. For more complex forms, the R:Base forms commands can be used for data entry or data editing.

Edit allows existing data to be edited in the data base by using forms defined either with the R:Base forms commands or by using the vertical form used for data entry.

MAIN MENU

Product Testing Lab
(1) UPDATE Records
(2) ADD Records
(3) REMOVE Old Records
(4) BROWSE through Data Base
(5) PRINT Reports
(6) ADD New Product Data

UPDATE MENU

Update Data Base
(1) Roster by Employee Number
(2) Roster by Employee Name
(3) Test Station Maintenance
(4) Product Test Data by Log No.

ADD MENU

Add New Records
(1) New Employee
(2) New Test Station

REMOVE MENU

Remove Old Data
(1) Old Employee
(2) Old Test Station
(3) Old Test Data by Date
(4) Old Test Data by Station

BROWSE MENU

Review Data
(1) Employee Data
(2) Test Station Data
(3) Product Test Data

PRINT REPORT MENU

Print/Display Reports
(1) Employee Data
(2) Test Station Data
(3) Product Tests by Product
(4) Product Tests by Station
(5) Product Tests by Operator
(6) Product Tests by Submitter

Figure 7.23   Menus designed for the product-testing laboratory application.

Delete allows the user to delete selected records from the data base.

Browse allows the user to view data in columns on the screen. If more columns of data are contained on a screen than can fit, the user can shift to the desired columns for viewing. Browse allows the displayed data to be edited. Browse also sorts as many as three columns of data.

Select is similar to browse, but the user can select to view one or more columns of data that meet specific criteria. Unlike browse, the displayed data cannot be edited nor can the displays be shifted to view more columns of data than can be displayed on the screen.

The *print* function allows data output to the printer or the screen by using a report format designed in application express. For more exotic reports, the R:Base report commands can be used.

To implement the menus, choose item 3, **Define a New Application**, from the application express main menu. Application express then displays the existing data bases and prompts the user to select one. Choose **TESTLAB** and name this application **TESTAPP**. Choose a vertical display type and enter the title and choices for the main menu as shown in Figure 7.23. Select the Esc key to exit the main menu. Then design a help screen that will be displayed when the user presses the F10 key.

Now the actions for each menu item in the main menu can be defined. Each of the main menu items moves the user to another menu except item 6. Select menu for the first menu item action. Now enter the UPDATE Records menu information. Select and enter the corresponding menu data for the other main menu items. At item 6, select the load action rather than a menu. This command will load data into the product test table that will be called ProdLog. This table and menu item will be the most widely used; that is why it was placed on the main menu. When application express prompts for the table to load, select the ProdLog table. Then a list of forms defined for this table will be shown, along with the option **New**. Select **New** because a new data entry form will be defined. When application express prompts for the name of the form, select **ProdLog**. The column names from the table are then shown on the screen so the user can choose which one to enter through this form. The user may select to enter only selected values or all values. For this example, select all. Now as directed, place prompts on the screen for each column. An input screen form has now been defined. When the load action is selected, the form will be displayed on the screen and the user will be able to enter the product test data.

The main menu actions have now been defined. Now define the actions for each of the submenu items. These actions are selected in the same way item 6 was defined from the main menu. Now define forms for editing, browsing, and selecting each submenu item. Continue entering the rest of the menu action items.

When the action definitions are completed, application express will write the application code in R:Base 5000 procedural language in ASCII and binary formats. The ASCII version can be modified with a text editor and recompiled if customization is desired on what application express has done. Perform manual modifications with caution, at least if application express is to be used for application modifications, because any manual modifications will be lost when application express is used on the old application.

To run the application, exit application express, select R:Base from the main menu, and select item 1, the R:Base command mode. The **R>** prompt, meaning R:Base is ready for a command, will be displayed. Type **RUN TESTAPP IN TESTLAB.APX** and press the Enter key. The file TESTLAB.APX is the name of the file containing the binary form of the application. On the screen, the application main menu will be displayed. Select item 2 from this first menu and enter some data into each table. Try out the other menu items to see how they work.

This application could have been implemented in other relational DBMSs such as dBase III or PC–File Relational. The mechanics of the implementation using the other programs will be different, but the strategy of having different tables of related data will be the same.

Products Mentioned in This Chapter

PC–File II (freeware,$25), PC–File III (freeware,$49) and PC–File Relational ($149): ButtonWare, Inc. P.O. Box 5786, Bellevue, WA 98006. (206) 746–4296 or the 24-h order line 1–800–J BUTTON

Reflex ($99), Borland International, 4585 Scotts Valley Drive, Scotts Valley, CA 95066. (408) 438–8696

Lotus 1–2–3 ($495), Lotus Development Corp, 55 Cambridge Pkwy., Cambridge, MA 02142. (617) 577–8500

1–2–3 Report Writer ($150), Lotus Development Corp, 55 Cambridge Pkwy., Cambridge, MA 02142. (617) 577–8500 dBase III ($695), Ashton–Tate, 10150 W. Jefferson Blvd., Culver City, CA 90230. (213) 204–5570

PC/Focus ($1595), Information Builders, 1250 Broadway, New York, NY 10001. (212) 736–4433

Paradox ($695), Ansa Software, 1301 Shoreway Rd., Suite 221, Belmont, CA 94002. (415) 595–4469

R:Base 5000 ($695), Microrim, Inc., 3380 146th Place, S.E., Bellevue, WA 98007. (206) 885–2000

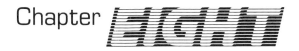

Project Management and Statistical Analysis Programs

aving projects completed on time and under budget is a major goal for many people. Projects come in all sizes, from the simple plans for a short business trip to major projects like the launching of a space shuttle. No matter how large or small, a project requires good management to be successful. Software for a PC can be a helpful tool for completing many project management tasks.

Project Management and Computers

Project management is not new. Even the most complex management aids can be performed with a pencil and paper. The key productivity asset of project management programs, like those found in spreadsheets and word processors, is the removal of mechanical drudgery from the task. Because of PCs, charts and diagrams can be assembled and printed in minutes instead of hours or days. Hundreds of tasks or activities can be coordinated; therefore, projects can be closely monitored and users can schedule resources to meet deadlines within projected costs. These programs are only tools, like hammers that can drive in nails. Having a good hammer does not ensure that the nail will be driven straight; you have to know how to use the tools.

The main strategies incorporated in project management programs are the critical-path method (CPM) and the priority evaluation and review technique (PERT). Both of these project management methods

1000–4/87/0187$06.00/1 © 1987 American Chemical Society

have been successfully practiced since the 1950s. Du Pont managers used CPM in the 1950s to plan and construct many of their chemical plants; PERT was first used in conjuction with the development of the Polaris submarine. Most project management programs also include the ability to produce Gantt charts and other visual aids. An example of a Gantt chart is shown in Figure 8.1.

Both CPM and PERT are based on a technique called *network analysis*. The network is a diagram showing the relationship between the separate tasks and activities that must be done to complete a project. *Activities*, or tasks, take time and resources to complete. When an activity has been completed, it is called an *event*. The box on page 189 shows the activities and events for a small project, building a fence. Some activities must be finished before others can be started. In the example of building a fence, the gate cannot be attached until the fence is complete, the fence boards cannot be nailed onto the frame until the frame is complete, the frame cannot be made until the posts are in place, and the posts cannot be set until the post holes are dug. Tasks like these are shown sequentially in a network diagram.

Other tasks may be concurrent. For example, while the gate is being hung, a first coat of paint can be put on the fence. These

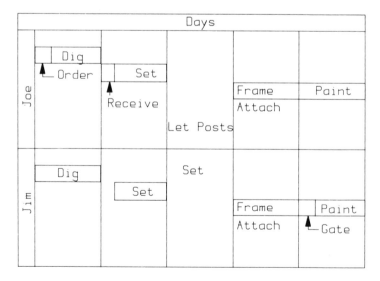

Figure 8.1   The fence project Gantt chart, generated using the Gantt Lab Manager from Gantt Systems.

Activities and Events for Erecting a Fence

Activity	Event	Activity	Event	Activity	Event
1 Order supplies	Supplies ordered				
2 Deliver supplies	Supplies received	3 Dig post holes	Post holes dug		
		4 Set fence posts	Fence posts set		
		5 Frame fence	Fence framed		
		6 Attach fence boards	Fence boards attached		
		7 Hang gate	Gate hung	8 Paint fence	Fence painted

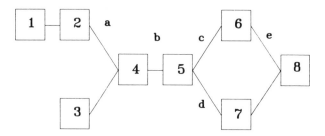

Figure 8.2 Network view of the fence project. Activities are shown as boxes connected by lines.

activities are shown on the same line as other, independent tasks but aligned with dependent tasks. Figure 8.2 shows a simple project network. The numbered boxes are events, and the lines between the events are activities. If the approximate length of time each activity will take and the available resources are included, the amount of time and resources that the project will take can be predicted. Each of the activities has a finite duration, so the length of the total project can be determined by adding the tasks.

The project shown in Figure 8.2 has two possible event sequences: a–b–c–e or a–b–d–e. These two sequences often require different amounts of time. The *critical path* is the sequence that takes the longest time. If the project is to be completed on time, the tasks and activities on the critical path must stay on schedule.

The duration of the project can be shortened by putting more resources into tasks on the critical path. If activities on the critical path are significantly shortened, though, another path may become critical by default.

Viewing the time phasing of a project is possible on a Gantt chart, which can be developed as shown in Figure 8.1. A Gantt chart shows which activities are being performed at what time.

Project Management Software

Most project management software allows the user to analyze a project by using one or more network analysis techniques. These programs allow the activities and their duration to be entered. The critical path will then be displayed, and network and Gantt charts can be produced. Some key features to look for in project management software are the following:

- ease of use (How easy is it to edit an activity? Do all commands have similar structure? How long does it take to learn?)
- logical links among pieces of data (Can the starting date or time of the project be changed and all other times updated without manual change?)
- graphs and plots (Can graphs be plotted on an *X–Y* plotter for presentations?)
- resources and cost (How important is resource allocation for the projects? How important are costs?)

One of the central themes of this book is showing how a PC can be used to turn data into information. As discussed previously, the data–information transformation requires more effort than simply plugging in a program and commanding it to analyze. The PC is a computing tool, not an answer machine. A PC can be provided with data to process, but the user must interpret the results.

Statistical software provides the user with computing tools that take disks full of data and turn them from meaningless numbers into information upon which to act. This software can help the user make meaningful consolidation and interpretation of raw data. The software gives the user the ability to view trends, make correlations, and perform regression analyses. These powerful computing capabilities are not accomplished by just plugging in a disk and typing **GO**. As with all new tools or instruments, statistical software must be tested on known sets of data to ensure correct operation of the program and correct assumptions from the program results. In other words, even powerful computing tools have no shortcuts, and therefore brainware is extremely important. Spend the time to learn the operation of the program and the meaning of the program's results.

Four main groups of abilities are needed for interpretation:

1. The ability to make computations, manipulate and transform data, correlate data with other values, and perform regressions. The types of programs that can do these functions are spreadsheet programs.

2. The ability to store and retrieve data. Data must be stored in a logical manner so the user can perform analyses. Stored, or historical, data can indicate the trends for the future. Calibration data must also be retrievable. Spreadsheets and data management programs have this ability.

3. The ability to plot graphs and display data. Graphics can easily show trends in data. The types of programs that have this ability

are spreadsheets, graphics programs, and some data management programs.

4. The ability to measure if an observation is statistically sound. Statistical programs and spreadsheets can do this with the proper equations.

Spreadsheets are contained in all four program categories. The many capabilities of spreadsheets have made them popular products. Spreadsheets are excellent at taking raw data and interpreting it into information. Some industry observers call these programs decision-support systems because they aid the user in making decisions. Many features found in stand-alone statistical products are being added to spreadsheet and data management products. For example, Lotus 1–2–3 version 2 includes regression analysis. Other new products like Reflex, Javelin, and Paradox provide similar computing and graphics capabilities but use a data base management system (DBMS) rather than a spreadsheet as their way to store data. Spreadsheet programs provide the capability to perform statistical calculations. Most of these programs provide the first level of calculations as built-in functions, such as calculation of mean, standard deviation, variance, minimums, and maximums. These programs provide the computing tools to help the user make decisions about data. So, although stand-alone statistical programs provide number-crunching capability, other programs that can also perform these tasks should not be ignored.

STATA and Statgraphics are stand-alone statistical programs. Both programs have their roots as mainframe statistical programs that have been written to run on the IBM PC and compatibles. Both programs can calculate means, standard deviations, and correlation coefficients and can perform single- and multivariable regressions. Both programs provide the ability to plot data in many formats. The differences in these programs involve how the user interacts with them. STATA is command-driven, and Statgraphics is menu-driven. The newest version of STATA has an optional menu module that is activated by pressing the F10 key. The user is then presented a list of commands that can be selected. STATA offers fewer preprogrammed functions, but the user can build any function needed with the necessary statistics. Statgraphics provides a large number of prewritten procedures; therefore, the user needs to know little about the mechanics of statistics to have a successful start. Of course, it takes time for the user to learn how the program operates and to have full confidence in the results.

STATA

STATA displays data in columns like a spreadsheet. The STATA commands allow the user to type in columns of data or import data from ASCII files. These commands also allow data to be manipulated by sorting, editing, labeling, and merging. The computing commands provide various functions. The command **SUMMARIZE** calculates means, standard deviations, variance, minimums, maximums, skewness, kurtosis, and percentile distributions. **CORRELATE** calculates a correlation matrix. **TABULATE** provides frequency and percentage summaries. **REGRESS** calculates linear multivariate regression coefficients, standard errors, t-statistics, ANOVA tables, and r^2 statistics. Polynomial regressions can also be performed.

STATA can make text character plots by using the vertical line, dash, or asterisk. If an add-on STATA/Graphics module is purchased, high-resolution black-and-white bar, pie, and scatter plots can be created. Scatter plots include regression lines. The plotting feature that seems most useful in this program is the scatter plot matrix of as many as eight variables. STATA/Graphics can divide the screen into a maximum of 64 separate graphs and plot the scatter plot correlation for every combination of eight variables. In this way, one screen of data can quickly tell the user which variables correlate and which are completely independent.

STSC Statgraphics

Statgraphics is the stand-alone statistics program for those who work with APL, as most statisticians do. Although Statgraphics runs much slower than STATA because the program runs as an interpreted APL program, the number of statistical and numerical methods available via menus is impressive. For example, a Gaussian 32-point numerical quadrature and Holt linear exponential smoothing are two of the 150 menu items in this system.

The program also provides many types of graphs including two- and three-dimensional scatter plots, surface plots, contour plots, two- and three-dimensional histograms, and pie charts. Data can be entered by importing ASCII, Data Interchange Format (DIF), and Lotus 1–2–3 files.

Products Mentioned in This Chapter

Gantt Lab Manager ($395), Gantt Systems, Inc., 495 Main Street, Metuchen, NJ 08840. (201) 494–7452

STATA ($395, $195 graphics add-on module), Computing Resource Center, 10801 National Blvd., Los Angeles, CA 90064. (213) 470–4341

Statgraphics ($695), STSC Inc., 2115 E. Jefferson St., Rockville, MD 20852. (301) 984–5000

Section THREE Communication and Interfacing

Knowledge is of two kinds. We know a subject ourselves, or we know where to find information upon it.

—Samuel Johnson

Information has two major components, accuracy and time. Good information has both; bad information lacks either or both. Scientists know this fact all too well. The software discussed thus far helps to produce accurate results. The software and hardware discussed in this section will aid you in getting the information to the right person at the right time.

Computers can send messages around the world to thousands of people. When the time element of information is critical, a quick, well-organized, and well-presented message is invaluable. An electronically transmitted message gets to its destination immediately and demonstrates that the sender is using the cutting-edge products of the technological revolution.

Exploring how a PC can communicate with other computers and devices is the goal of this section. Data communication for a PC system is a two-way highway. In-bound data is collected from physical phenomena-measuring transducers, monitored from dedicated computers imbedded in instruments, and received from computerized data bases or through one-on-one communication with other scientists through networks or computer–computer links. The results of the analysis are outbound data: reports, charts, graphs, and tables generated from the captured data. If experiment control is done, the experimental control signals may depend on the results of acquired data. In this section, how a PC performs these communications tasks will be discussed.

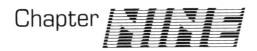

Data Communication Interfaces

T̄ransmitting and capturing data to and from other sources requires some physical links, or *interfaces*, to the real world. In most cases, these interfaces do not require knowledge about how they work and can be used as black boxes that perform specific tasks. Normally, these interfaces are supplied in the form of printed circuit boards, which are inserted directly into the PC expansion slots, or external boxes, which communicate through RS–232C or another communications link. For full operation, simply insert or connect them to the computer and execute the proper software. The difficult task is finding the proper interface for the desired application and then knowing how to verify correct operation of the interface.

For data communication between computers, interfaces are relatively standard because both computers must use the same interface and communications method. The most common computer–computer interface is RS–232C. Using an RS–232C interface and modem, the user can communicate with a computer in the room next door or on the other side of the world. Because most laboratory instruments have internal computers, the user can communicate with the instruments if the instruments have the ability to communicate. This chapter will describe the most common computer–computer interfacing methods, including computer–instrument interfacing and computer–computer communication via networking.

Asynchronous versus Synchronous Communication

Asynchronous and synchronous are terms used to describe the timing of the transmission and the receipt of communicated data. These

1000–4/87/0197$07.50/1 © 1987 American Chemical Society

terms describe methods by which the sending and receiving computers extract meaningful bits of information from the steady stream of data each computer sends or receives.

When information is transmitted by the asynchronous method, the sending computer injects special characters, called *start* and *stop bits*, to the data stream that indicate where a character's data begins and ends. These special bits "frame" the actual data bits. The start bit is always a binary 0. This binary 0 is recognized as a change from the resting line-signal voltage representing a binary 1. After the start bit, the actual data bits follow and provide the pattern of 0s and 1s that represent single characters. Stop bits are binary 1s that mark time until another 0 arrives. Using this method, data can be sent at any time; no problems are caused by slow or fast telephone connections. *Protocol* is a formalized set of conventions used for establishing and maintaining contact between two communicating devices. The most common asynchronous hardware protocol is RS–232C. Common file transfer protocols include XMODEM and the Kermit protocol. These protocols include error checking to ensure proper communication.

Synchronous communication describes a method of continuous data transmission in which blocks of data are transmitted without marking each individually transmitted character with start and stop bits. Instead, the receiving computer uses a very accurate clock to select where to divide the data stream into individual characters. Each block of data also includes error-checking information, so transmission errors can be detected. This precisely timed method is usually used only for communication between mainframe computers and their terminals. The most common protocols used by IBM mainframes are Bisync (BSC), synchronous data link control (SDLC), and high-level data link control (HDLC). These protocols establish the wiring, handshaking, and error checking necessary for data communication. Older products use BSC, and newer products use SDLC.

Serial versus Parallel Communication

The essence of communication between computers is simply the sending and receiving of bits, 0s and 1s. Using ASCII codes, sending one character requires sending and receiving 8 bits. Two obvious methods can be used to send these 8 bits. One *wire*, which is an electrical path along which data is transmitted, could be used and each bit sent one after the other in *serial* fashion. Once 8 bits are sent and received, one character is transferred. The other method would be to have eight separate wires, each carrying one bit. A character using this method would be sent together in parallel fashion. Using this method, data is transferred eight times faster but requires eight separate wires.

Both methods of communication are used. Serial communication is used most commonly to communicate between remote computers because telephone lines are used, and telephones have one-signal lines. The parallel method is used between the PC and most printers. Here an eight-wire cable is required, and the short distance and faster printing speed enhance the computer system.

The following interfaces are used to send and receive digital data from one computer or instrument to another. Each interface has its own method of data transmission and communications control protocol.

Asynchronous Communication: RS–232C and Modems

Before a computer can use asynchronous communication with another computer, both computers need three items: a modem, an asynchronous communications (serial or RS–232C) port or board, and communications software.

The Modem

A *modem* is a device that links a computer to the telephone line and allows the computer to communicate with a remote computer. The term *modem* is a contraction of the terms that describe its two major functions, modulation and demodulation. *Modulation* is the process of converting the digital data (bits) stored in a computer into analog (voice frequency signals); in this way the data can be transmitted via telephone lines. *Demodulation* is the process of converting the analog signals back to digital data. Modems can be placed either internal or external to a computer. Internal versions plug right into an expansion slot and have a common telephone plug facing the outside of the computer to make the telephone connection. External versions require an asynchronous communications port in the computer that connects to the modem, which in turn connects to the telephone line. Modems differ in their transmission speeds, and a number of special features can be added, including telephone automatic dialing and answering.

The Asynchronous Communications Port

Asynchronous communication is also called serial communication, or RS–232C. RS–232C is the rule number that determines the method of communication. The rule number is set by the Institute of Electrical and Electronic Engineers (IEEE).

The hardware links used in serial communications are cables (Figure 9.1). Each cable contains as many as 25 wires. Normally, only 10 or fewer wires are used as signal or control lines. These cables are attached to connectors on printed circuit boards, or cards, inserted in an open expansion slot in the IBM PC. Asynchronous ports can be purchased as a card or bundled together on a board with four or five other pieces of hardware, such as memory, parallel printer ports, joy stick ports, and clock–calendars. These multifunctional cards are economical when the number of expansion slots and the price per function are considered.

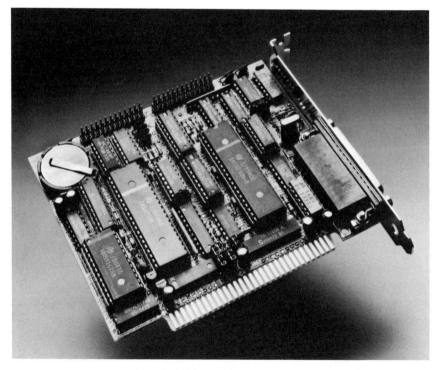

Figure 9.1   RS–232C serial communications interface for the IBM PC with cable and connector. This interface card is inserted into one of the expansion slots of the IBM PC.

The asynchronous communications method uses the same strategy used by the earliest electronic communications devices such as the telegraph. One signal wire can transmit one of two voltages representing a bit, either a 0 or 1. Each character sent is composed of a specific combination of 0s or 1s.

Contrast this method of communication with parallel communication, which transfers 8 or more data bits simultaneously. Each type of data transmission has unique advantages and disadvantages. Serial communication, which sends only one bit at a time, is simple to implement and requires only one wire for data and one for ground. It is also perfectly suited for transmitting data over long distances by existing telephone systems.

In asynchronous communication, groups of bits are transmitted with an arbitrary length of time between them. Figure 9.2 shows a typical asynchronous transmission of characters.

Figure 9.3 shows the transmission of the character *G* using ASCII code. The data is transmitted over the link by two voltage levels that represent the two possible states of a binary digit. A logic 0 (SPACE or high) is represented by a voltage range from +3 to +5 V, and a logic 1 (MARK or low) has the range from –3 to –25 V. The transmitting

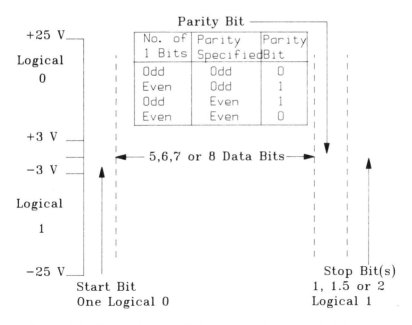

Figure 9.2   Transmission of characters via an RS–232C serial interface uses two voltage levels to represent either a 0 or a 1.

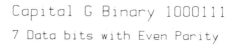

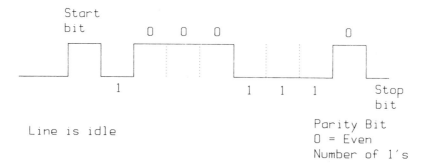

Figure 9.3   Transmitting the character G using RS–232C and ASCII code is similar to sending the character over telegraph via Morse code.

device sends data by placing the line in the high state for a 1-bit time. This first bit is the start bit. Following the start bit, the data bits are sent followed by an optional parity bit and 1 or 2 stop bits. The speed of a transmission is set in bits audible, or *baud*. Both the transmitting and receiving devices must use the same transmitting and receiving protocol; otherwise communication will not be completed.

Data Communication via RS–232C.   In any data communication session, two computers are involved. The transmission of data can occur in one of three ways: in one direction (*simplex*), in two directions but at different times (*half duplex*), or in both directions at the same time (*full duplex*). These configurations must be set by the user.

   The simplex configuration allows data transfer in only one direction. Just as cars and trucks can only go one way on a one-way street, in simplex configuration, data can only travel in one direction. When the communications link is established, one computer is designated as the sender and the other as the receiver. The sender can only send data and not receive, and the receiver can only receive data and not send.

   The half-duplex configuration allows each computer to send and receive data, but communication can only occur in one direction at a time. This situation is like allowing traffic to go only north on a street in the morning and then south on the street in the afternoon. Traffic can go in both directions, but only in one direction at a time.

The full-duplex configuration allows each computer to send and receive data in either direction at the same time. Data can be sent and received at the same time. This situation is like that on a two-way street. Traffic can go in both directions at the same time. In full-duplex mode, both computers can both send and receive data, but one computer usually sends data while the other receives. The receiving computer echoes the transmission it receives back to the the sending computer to confirm accurate transmission of the data. Thus, the sending computer actually sends and receives the same data, checks for error, and then continues to send data. The configuration of both the sending and receiving computers must be the same for proper communication.

The number of data bits is the number of bits that actually form the character. This does not include any start, parity, or stop bits. ASCII character codes are made of seven characters. For example, the bit code for the letter *R* is 1010010. Other character codes may be made of 5, 6, or 8 bits.

Parity provides a method for error checking the communication. The parity bit is optional and can be set to either *1* or *0*, which performs no error checking. Setting the parity bit to odd or even provides error checking by counting the number of *1* bits in the data bits. For example, if the number of *1* bits in a character is odd and the parity specified is also odd, then the parity value is *0* (Figure 9.2). If the number of *1* bits in a character is even and the parity specified is also even, then the parity value is *0* again. When the number of *1* bits in a character and the parity specified are odd and even or even and odd, then the parity value is *1*.

As another example, the letter *R* (1010010) contains three *1* bits. If odd parity was specified, the parity bit is *0*. If even parity was set, the parity bit is *1*.

Stop bits are not really bits at all. The transmitter holds the line in the idle state for the number of bit times specified by this parameter.

Baud rate is the number of bits per second transmitted. There is a relationship between baud rate and the maximum distance for the cable. Twenty feet is about the longest distance that can be used if the operation is set at 9600 baud by using a ±12-V signal level. But if a slower baud rate of 300 is used, 500 ft of cable can be used with no problems. The voltage levels and cable quality are involved in the success of the communications, so the system should always be tested under the conditions of operation.

Handshaking is used to communicate device-status information from one device to the other. These handshakes can indicate a buffer-full condition, a receive data errors condition, modem status, and a ready-to-receive condition. Examples of handshaking protocols are enquire/acknowledge (ENQ/ACK) and XON/XOFF.

ENQ/ACK handshaking is used by some serial devices to detect a buffer-full, or a not ready, condition. When the transmitting device sends a line of text to a receiver, an enquire character (ASCII 5) is also transmitted following the text. The transmitting device waits for an acknowledge (ASCII 6) from the receiver before the next line is sent.

XON/XOFF, also known as DC1/DC3, is another handshake method. During data transfer, the receiver monitors its input buffers to ensure sufficient space for at least one more line of data. When sufficient space is not available, the receiver sends an XOFF to the transmitter. The transmitter then suspends further transmission until the receiver sends an XON, indicating that the transmission can resume.

File Transfer Protocols. File transfer protocols use hardware handshaking to transmit and receive whole files of data. Examples of file transfer protocols that use handshaking and retransmit data that was not received correctly are the XMODEM and Kermit protocols.

XMODEM protocol refers to a commonly used, public domain, error-checking protocol. Under XMODEM protocol, information is transmitted and received in 128-byte blocks. After each block is received by the computer, the block is checked for accuracy of transmission. Blocks of data not correctly received are retransmitted until no errors are detected. Both the transmitting and receiving computers must be running a program to perform XMODEM protocol. Fortunately, most networks and bulletin board systems use this protocol; it is integrated into most communications programs that run on a PC, such as PC–TALK and QMODEM. Both PC–TALK and QMODEM are freeware products. XMODEM protocol is used to transmit binary code files such as executable (.EXE) and .COM programs and is excellent for transmitting files from one PC to another.

Checksums are used to detect errors in transmission of data. After each block of data is transmitted, an additional value, called the checksum, is sent. The checksum is calculated by adding the ASCII values of each character in the 128-byte block and dividing by 255.

After each block is transferred, the receiving computer calculates its checksum and compares the result with the checksum received from the transmitting computer. If the two values are the same, the receiving computer sends an acknowledge (ACK) character to tell the transmitting computer to send another block. If the two checksum values are not the same, the receiving computer sends a not acknowledged (NAK) character to request a retransmission of the last block. This retransmission process is repeated until the block of data is properly received or until nine attempts have been made to transmit a block. If the block cannot be properly transmitted after nine attempts, the data transmission will be aborted.

If files need to be moved from one type of computer to another, for example, from a DEC–VAX to an IBM PC, or from a PC to an IBM 370, then Kermit protocol should be considered. The Kermit protocol uses a strategy similar to XMODEM to detect errors in transmitted blocks, or packets, of data. The Kermit protocol was developed at Columbia University to allow data file transfers from different computers. Programs that use the Kermit protocol have been written for many diverse computers, including the IBM 370 series running the VM/CMS operating system; VAX–11 running VMS, VAX, or Sun; PDP–11 running UNIX; DECsystem–10 or DECsystem–20 running TOPS–10 or TOPS–20; PDP–11 running RT–11, RSX, or RSTS; Intel–8080, Intel–8085, or Z80-based computers running CP/M–80; Intel–8088, Intel–8086, or Intel–80286 running PC–DOS or MS–DOS; and Apple II running Apple DOS.

For the Kermit protocol to work, both computers must run the Kermit protocol program. Files can then be transferred from the transmitting computer to the receiving computer. The Kermit programs are an excellent way for different computers to share data. Because it was developed using government grants, Kermit is available as public domain software. Columbia University is willing to provide all Kermit programs, sources, manuals, and other documentation to academic or corporate computing centers in return for a modest fee to cover the cost for media, printing, postage, and labor. An excellent set of articles describing the Kermit protocol by Frank Da Cruz and Bill Catchings appeared in BYTE magazine in June 1984 (page 255) and July 1984 (page 143).

Serial Hardware Connections

Unfortunately, because the RS–232C standards were loosely written, some vendors of communications equipment have interpreted the standard differently. Be aware that on IBM PCs and most PC clones, the parallel printer port uses the same DB–25 pin connector. Be sure when you are setting up the computer that you know which cards have a parallel port and which cards have a serial port. Not knowing the types of ports leads to frustration when two devices must be interfaced; both devices may have RS–232C interfaces but may not connect correctly. Two standards for wiring the interfaces are data terminal equipment (DTE) and data communications equipment (DCE). Table 9.1 shows some pin assignments on a 25-pin connector. Along with different pin assignments, different connector types might be encountered. Complete trouble-shooting for communication between two RS–232C-equipped devices is beyond the scope of this book. Two excellent books can be consulted to provide step-by-step instructions for connecting two devices via RS–232C (see end of chapter).

General-Purpose Interface Bus

A *general-purpose interface bus* (GPIB) was designed to provide mechanical, electrical, timing, and data compatibilities among all devices adhering to the standard. GPIB is also called IEEE–488 and HPIB. IEEE–488 is the rule number for this communications method. HPIB is the abbreviation for Hewlett–Packard Interface Bus, the Hewlett–Packard implementation of the IEEE–488 standard. In essence, interfacing an IBM PC to other devices has been simplified by using

Table 9.1   Some Pin Assignments on a 25-Pin Connector

Cable Pin	Signal Name	Function
1	Ground frame	Output data for DTE
2	Transmit data	Input data for DTE
3	Receive data	Input data detected by DTE
4	Request to send	Indicates to DCE that DTE is ready to transmit
5	Clear to send	Indicates to DTE that DCE is ready to accept data
6	Data set ready	Indicates that DCE is not in test mode and power is on
7	Signal ground	
8	Data carrier detect	DCE is receiving carrier signals
20	Data terminal ready	Tells DCE that device is ready to transfer data
23	Data signal rate	Selects one of two rates on a two-speed modem

Figure 9.4   A GPIB interface cable. Cables can be stacked on each other; additional devices can then be connected.

a GPIB. You do not have to worry about how to make the connections between the compatible devices; you need only consider what the devices will communicate. All devices are linked together by two-headed cables like those shown in Figure 9.4. These cables are attached with one end on the GPIB device, of which there can be as many as 15; the other end is attached to a connector on a card in one of the expansion slots.

For conceptual purposes, the organization of GPIB can be compared with that of a committee, with all "members" strictly adhering to the rules. Each member of the GPIB is identified by a unique number that is usually set by flipping a set of dip switches. One member is designated the "chairman", "master", or system controller. Normally, the IBM PC is the system controller, or master. All other devices wanted for communication must be talkers–listeners and not controllers. On a committee, only one person at a time can speak, and the chairman is responsible for designating who can speak. On a GPIB, only the master can designate which device can "speak"; therefore, only one master is allowed. The GPIB device allowed to speak is called the active talker. As the active talker sends a message, the other devices on the bus have the opportunity to listen. The master tells which devices should listen to the message.

When a GPIB interface card like the Ziatech ZT–1444 is purchased, software support should be received to perform communication through the interface. Ziatech provides or offers this communications software to perform the interactions. These programs are written in assembly language and can be called from programs written in BASIC, Compiled BASIC, Pascal, or C.

Some BASIC statement lines that communicate on the GPIB include BASIC command (**CALL ABSOLUTE**), device to which to send message (12), message to send (I), and location of assembly language routine (SENDS%). These statement lines are written as **CALL ABSOLUTE ("12", "I", SENDS%)**. This last command sends the letter I to the device numbered 12 on the GPIB network. To set up this capability, first an assembly language program called ALL.MB should be loaded into memory with the command **BLOAD "ALL.MB"**. When the **ALL.MB** code is run, variable names such as SENDS% and RECVS% are assigned values that correspond to the starting location for the command codes. The **CALL ABSOLUTE** command can then be used to execute the portion of the assembly language program that performs the desired action. The parameters to use for the actions such as the "message or command to send" and the "device number to send or listen to" is passed to the assembly language program as part of the **CALL ABSOLUTE** command.

Similarly, **CALL ABSOLUTE("8",NEWDATA$,RECVS%)** is used to receive data from device 8 on the GPIB network and place the data in the variable string named **NEWDATA$**.

GPIB network status or errors set by devices on the network can be collected with commands like **CALL ABSOLUTE(STATUS%, ERRORST%)**. This command returns a value for the variable STATUS%, and the value can be checked against a table of error status conditions defined by the GPIB card vendor.

Binary Coded Decimal

Binary coded decimal (BCD) is a method of encoding the decimal digits (0–9) in a 4-bit format. Table 9.2 shows the binary coding of the 10 decimal digits and the six additional ASCII characters that are

Table 9.2   BCD Coding

ASCII	Positive Signal as Binary 1	Negative Signal as Binary 1
0	0000	1111
1	0001	1110
2	0010	1101
3	0011	1100
4	0100	1011
5	0101	1010
6	0110	1001
7	0111	1000
8	1000	0111
9	1001	0110
:	1010	0101
;	1011	0100
¼	1100	0011
=	1101	0010
½	1110	0001
?	1111	0000

allowed. One of two formats is selected; either the negative signal is designated as binary 1, or the positive signal is designated as binary 1.

Not as many instruments use BCD as an interface as in the past. Some instruments that still use BCD are analytical balances and automatic chromatography injection samplers.

BCD data can be read by most multiline digital *input/output (I/O) boards*. Unlike the serial data transmission previously discussed, in which only one wire carries the data, BCD requires multiple lines. Four lines must be used for each character, and the values of 0 or 1 are read from all lines at one time. If multiple digits must be sent, either more sets of four lines must be used or a method of synchronizing the transmission of characters, called *strobing*, must be implemented. Normally, a fifth strobe, or data-ready line, is used for this purpose.

A number of multiline digital I/O boards are available. A list of vendors of these boards is provided at the end of this chapter. Most boards include the source code software necessary to read these ports, but the user has to write the software necessary to decode the 4-bit clusters into decimal values.

Capturing Data from Dedicated Instrument Data Systems

Today, many laboratory instruments can send the results of their analysis to a computer. These instruments usually have their own proprietary data acquisition, data reduction, and sometimes graphical data display capabilities bundled with instrument and experiment control. Normally, the final results of an analysis are sent via serial interface (RS–232C), GPIB, or BCD. Unfortunately, most of these instruments are not able to save results for later transmission. Usually, after each analysis, the data is immediately sent to the communications port. Capturing this data requires the user to dedicate a PC to data acquisition, operate one of the concurrent or multitasking operating environments like Microsoft's Windows and dedicate a window to the acquisition, or have a device that buffers or saves the data as it is sent.

Dedicated PCs

Dedicated PCs are the simplest way to acquire data from an instrument. With the cost of an IBM PC or a PC compatible now well below $1000, this solution is not expensive. If the same instrument requires additional control, then dedicating a PC for real-time control and data acquisition may be required. The same PC can be included in a local area network (LAN) for a few hundred dollars; the acquired data can be accessible to everyone on the network.

If the instrument has an RS–232C interface, then the proper hardware connections must be made before data can be captured by a *terminal emulation program*. A terminal emulation program allows the user to set all the necessary communications parameters and start communication. Incoming data can be saved on disk for further data analysis.

If the instrument has a GPIB interface, the user must write the software necessary to communicate and capture data built around the software routines supplied by the GPIB board vendor. All GPIB board vendors use different codes to communicate with their own boards. This means the software to acquire data via a GPIB must match the GPIB card used.

If the instrument has a BCD interface, the user must write the software necessary to communicate and capture data built around the software routines supplied by the interface manufacturer. Each digital I/O interface manufacturer uses a different code to communicate with the board.

Most interfacing to instruments requires a custom job for each instrument. As more instrument manufacturers realize the utility of communication with an IBM PC or PC compatible, more of them will provide the ability to communicate with a PC via standard communications ports. Many instrument manufacturers already have products that can use an IBM PC for data acquisition, reduction, storage, and instrument control.

Concurrent Windows or Multitasking Solution

This strategy uses the same programs just described. The major difference is that as data acquisition and control occur in one task, or window, other applications can be performed in other windows. This strategy is excellent if it works. Unfortunately, the concurrent windowing and multitasking environments may not be supported by the other software desired. Each user must experiment to see if the desired programs are compatible.

Buffering Data Solution

A third low-cost solution to this problem is use of one or more printer buffers. Printer buffers are relatively low-cost ($200–$500) depending on the amount of memory (8K–512K). Printer buffers can capture the data as it is transmitted in real time. Then, when the IBM PC is ready for the data, the user can read in the data from each data buffer by simply pressing the copy button on the buffer's front panel or by electronically signaling the interface by setting the proper handshake lines to the correct position.

A relatively small program can capture data from the printer buffer and then store the data on disk for further data processing. Nelson Analytical offers a set of programs called CAPTURE! that uses this strategy. The main program handles the interaction with the printer buffers, and smaller "filter" programs control the storage of data on disk. The filter programs store the data on disk in the most convenient format for processing the data. The captured data can then be analyzed by other programs.

PC–Mainframe Connections

Advertisements for PC–mainframe link products make communicating with a mainframe seem as easy as loading new software or plugging in a terminal emulation board. This simplicity is not always the case. More energy may be required to send or receive data from the corporate mainframe via a PC than would be required to rekey data from printed reports. To operate efficiently, links must be implemented with the cooperation of both the end user and the mainframe data processing staff. Usually, one or more of the data processing staff acts as a liaison to the PC users because most mainframe software is not designed for casual users.

Linking PCs to mainframe or minicomputers requires both hardware and software. A physical link with the host computer and at least two programs are required, one program for the PC, and one program for the host computer to perform the communication.

Two modes of communication, asynchronous and synchronous, are common with the host. Asynchronous communication has already been discussed and will be discussed further in the next chapter. Minicomputers such as the DEC–VAX can be accessed by simple asynchronous communication requiring only software emulation.

Hardware

Synchronous hardware links are used by IBM mainframes. Figure 9.5 shows a typical synchronous communications configuration. The host computer is a large computer containing data and programs for a whole organization. It could have billions of bytes of disk space and

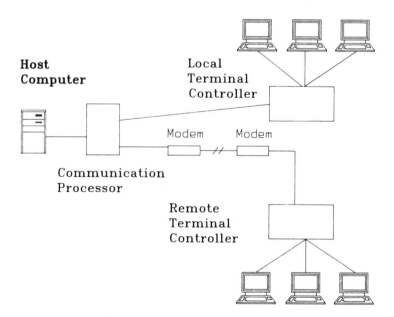

Figure 9.5 Communication with an IBM mainframe computer. The host computer can serve thousands of users via terminals in the same building or in remote sites.

millions of bytes of random access memory, high speed printers, and hundreds of terminals to enter and retrieve data. The *communications processor* is a second computer designed to control communication between the host and the terminals. The communications processor is located a short distance from the host computer and is connected to the host via a high-speed communications line. The terminal controllers can be local and connected via cables or remote and connected via synchronous modems.

For a PC–mainframe link to operate, the regular synchronous mode of data transmission must be emulated on the PC. This emulation is usually done by using special-function, printed circuit boards that are added to the expansion slots of the PC. These boards make the IBM PC, or PC compatible, act like a normal IBM display terminal. These products usually emulate IBM's two most popular terminals, the 3278 and 3279. The user can toggle (switch) back and forth between terminal emulation mode and stand-alone PC–DOS mode. In terminal emulation mode, for example, the PC has all the features of the 3278/79 terminals—status-line display, full-screen format, and two- and seven-color modes. Most vendors also include file-transfer capabilities, a feature not available on the 3278 or 3279 because these terminals do not have disk drives.

The leading board for this application is the IRMA board by Digital Communications Associates. The IRMA board and software enable a PC, or any other ASCII device, to connect via coaxial cable to a controller, which in turn connects to the mainframe. Directly connected cables transmit data many times faster than modems, but for remote communications, access via modem is the only possibility. For remote PCs, DCA offers IrmaLine, an interface box installed at the controller. Remote PCs dial IrmaLine, which then connects the PC to the mainframe through the controller.

Software

Once the hardware link is obtained, the proper software must be acquired to perform communication. Software that runs on the PC and also on the host computer is required. Most mainframe and minicomputers do not currently have the proper software to handle this type of communication. This lack of mainframe software is one of the most hotly debated issues in many data processing (DP) and management information services (MIS) departments of many companies. Understandably, DP or MIS managers would rather not allow PC users to have free access to their computer resources if valuable or sensitive information is stored and processed in the mainframe. In addition, the software necessary to communicate intelligently with PCs and allow PCs to use the mainframe data must be written. This additional job reduces the DP or MIS department's resources.

A leader in providing linked software products is Informatics General, which provides the Answer series, an open-ended link that allows data processing to maintain overall control but provides users powerful, menu-driven report-generator capabilities. The product consists of two parts. Answer/DB extracts data from almost any IBM data base and transmits the data to the PC via software in the PC-like Lotus/Answer, which receives the data and generates a Lotus 1-2-3 or Symphony file. Informatics General also offers dBase/Answer, which downloads data into Ashton–Tate's dBase II or dBase III. The Answer series offers users uploading, but not updating, capabilities to keep mainframe data from being destroyed. Uploaded data goes into a special mainframe file set up by the mainframe staff. This data can then be integrated into the proper data bases by the data processing staff.

Terminal Emulation

Luckily, alternative ways are available to communicate between PCs and host computers without any special software running on the mainframe. Because mainframe and minicomputers are designed to communicate with users through terminals, a way to link a PC with the host computer is to have the PC emulate a terminal.

A PC, through software and hardware, has the ability to imitate almost any type of cathode ray tube (CRT) terminal, graphics terminal, cluster controller, or printer. The three main types of emulation include asynchronous, synchronous, and a combination of both. Asynchronous CRT emulations are the simplest and usually require only software, an asynchronous serial port (RS–232C), and a direct hard-wired connection or modem. This type of communication is simple, inexpensive, and slow. Normal rates are from 300 to 1200 baud. Some of the terminals emulated by these products are IBM 3101, DEC VT–52/100, HP–2621, Lear–Siegler ADM–3A, Data General D100, Televideo 912, and Tektronix 4010. These emulation products sell for $100–$300 and can replace a terminal that normally costs $1000–$2000. This savings and the convenience of having a PC convince many users who would have purchased a terminal to purchase a PC and a terminal emulator. Most of the Kermit programs also include terminal emulators for the other computer that the user plans to connect to for transferring files.

Synchronous emulations are complex because regular telephone lines are not adequate for the transmission rates and low noise requirements. This type of transmission requires a printed circuit card and emulation software. Synchronous transmissions move characters in a tightly ordered sequence. Encoding and decoding are based on precise timing of characters at the beginning of each data message. In most cases, the communications products for this type of inter-action are bundled together with the software to emulate a number of different terminals along with synchronous and asynchronous communication and a high-speed modem. These products cost from $1000 to $1500 and may require special telephone lines to get noise-free, high-speed (4800–9500 baud) interaction. Devices emulated by these products include the IBM 3278 and 3279 display terminal when attached to a 3274-type terminal controller, the IBM 3780 and 2780 or 2770 and 3770, and remote job entry (RJE) intelligent work stations.

Using either type of terminal emulation product allows the user to perform tasks on the mainframe computer. This capability gives the user access through a PC to all of the computing resources attached to the mainframe computer.

Although hundreds of communications products are available for PCs that provide the necessary protocols and procedures, most products work with only one type of computer and operating system. Some packages even require a specific brand of modem. Another potential stumbling block is that two character codes convert characters into binary code and vice versa. The character code most often used is ASCII. Almost all PCs used for communication use ASCII. However, all IBM mainframes use extended binary coded decimal information code (EBCDIC). So, for communications between an ASCII-based PC and an EBCDIC-based IBM mainframe, one of the systems must perform the data translation.

The next step toward full computer-networking capability is passing data from one computer to the other.

Most terminal emulation packages provide for some type of data file transfer. Some packages simply allow the user to invoke a text editor on the host computer to send a file from the microcomputer's disk to the mainframe as if the user were quickly typing the characters. Other packages automatically send the necessary log-on messages and enter account numbers and passwords needed to access a mainframe computer and a specific data base. Capturing data from the mainframe data base is done by opening a file and redirecting the characters that normally would go to the terminal screen into the file. These types of transfers work only with text files, and in most cases more data will be received from the mainframe data bases than is needed. The user must then write programs to select the data required. Selecting a subset of the data on the mainframe's data base or updating values in the mainframe data requires additional programming on the mainframe.

Looking to the Future

The intense interest in PC–mainframe communication has spurred the development of interesting new products that will most likely be future work stations. Examples of this new type of product include IBM's 3270 PC and the PC–AT/370. The 3270 PC can function as a standard IBM PC or as a high-resolution color terminal for several of IBM's mainframe computers. It can also perform both tasks simultaneously. As a result, the 3270 PC brings the utility of a stand-alone desktop computer along with the raw computing power and storage capabilities of existing mainframe computers. A special graphics version of the 3270 was recently announced with the capability to perform high-resolution graphics. This graphics version is a perfect hardware match for scientific modeling and data analysis. Now only software is necessary.

The PC–AT/370 can perform three distinctly different tasks. It can perform as a regular IBM PC and run PC programs, act as a terminal when attached to other host computers, and download programs written under the System/370 VM/CMS (virtual machine/conversional monitor system) operating system and run these programs without further interaction with the mainframe. Users can also write programs, files, and reports and send them back to the host computer for processing or storage.

The technology necessary to combine the strengths of PCs and mainframe computers is in its infancy. How these different resources are ultimately used in an organization requires careful analysis. PCs will have to be intelligently managed as information-processing resources and fit into an overall information network. Currently, no clear-cut choices are available to form the information network, and until the designers of mainframe and PC networks advance toward the same goals, users may not have a single product solution. Until the ultimate solution arrives, using the productivity products available will prepare PC users and their organizations for the future.

Local Area Networks for Scientific and Engineering Laboratories

Another method of communication between computers, called *networking*, has been developed to share data and computing resources such as communications links with other mainframe and minicomputers, printers, and plotters. When the computers are near each other, the network is called a *local area network* (LAN). LANs are experiencing explosive growth due to the large number of PCs that are in use and the need for users to communicate with other PC users and computers.

Advantages of a LAN in the Laboratory

A typical laboratory is well suited to take full advantage of local area networking. The LAN should be considered an important link in the four-level hierarchy of computers that a laboratory will use. At the first level, PCs can be used to acquire data, control experiments, analyze data, and generate reports. Each PC is a self-contained unit capable of performing all tasks independent of any other computing resource. Even if other computers in the hierarchy fail, or if users are unavailable, the individual PC user can still perform the task at hand. The PC user has a huge advantage over a user of a central computing resource because the PC user can have local data storage, data processing, and real-time experiment control. Also, the PC user has all the benefits a terminal user has on a larger computer if the PC user runs terminal emulation software.

At the next level, the LAN allows the sharing of experimental results within the laboratory. Now, reports can include data from many PCs combined without rekeying. The LAN can contain a laboratory-wide data base accessible by all PCs. Expensive peripherals such as letter-quality printers and *X–Y* plotters can be shared by each user on the LAN.

Summary reports that reflect the work from the entire laboratory can be sent to the next level of computers. These computers could be minicomputers, such as the DEC–VAX or IBM 4300 series computers, where the information from many laboratories is summarized and reported. Finally, the data is sent to the highest level, the corporate mainframe.

The LAN thus provides an efficient communication link between each of the PCs in the laboratory. This type of communication can be performed at a much lower cost but with less convenience by either using floppy disks to transport data from one PC to another or by using RS–232C communication from one PC to another. The RS–232C solution requires that both PCs be dedicated to the data transfer while data is being exchanged. With the expected growth of LANs, the current $1000 per PC for the LAN hardware and software should be reduced substantially. If the price can be lowered to about $200, then LANs will be a cost-effective method of data communication. The hardware needed to operate a LAN may soon be placed directly on the PC mother board just as the keyboard connector is connected to the mother board. This hardware will then be integrated into the price of each PC.

With IBM's introduction of their Token Ring network, most LAN observers agree that a significant standard has been implemented. In the coming months, many new LAN products will be developed, and the use of LANs in laboratories should become a reality.

LAN Software

Many major companies offer LANs and LAN products. Some of the most widely recognized companies include IBM, Corvus, 3COM, Nestar, Ungerman–Bass, Novell, AST, Quadram, Orchid Technology, and Fox Research. Features to consider in a LAN include LAN software, availability of application software, LAN media, and LAN topology.

The primary factors in selecting a LAN are the features and performance of the LAN software. LAN software should be thought of as an operating environment, just like DOS. The LAN software will provide additional functions to use or share with other devices such as disk drives and printers. The same software may also place restrictions on using common DOS functions. For example, the IBM PC network software does not allow use of the print screen (PrtSc) function, which immediately prints the contents of a monitor's screen.

The major features in the LAN software should reflect the intended use of the LAN and the LAN users. The software should be simple enough to operate so the users can get their work done even if they have had little training or have not used the LAN for a few weeks. Most LAN software thus uses a simple menu-driven setup and operation. For more advanced users, menus can be tedious, so ways to enter LAN setup parameters quickly should be available without going through numerous menus.

Application Software Performance on a LAN

Just because software runs on a single-user PC does not mean it can perform effectively on a LAN. In fact, unless the application's software vendor specifically lists networking capabilities, no assumptions about the performance of a specific application software should be made. In particular, DBMSs must be modified to run correctly on a LAN, otherwise the data in the data base can be corrupted, or made inaccurate. Most single-user applications are currently making the necessary changes to their packages so they will run well on a LAN. For example, both the popular data management programs R:Base 5000 from Microrim and dBase III from Ashton–Tate have single-user versions and LAN versions for the IBM PC-supported networks. The LAN versions of these data management programs can perform file and record *locking,* which means that only one user on the network can make changes to a particular record. All other users are locked out from trying to change a record the moment the other user is performing the change. This procedure ensures that each data item in the data base is changed by only one user at a time. Along with making the necessary changes to their products, the software vendors must also choose a LAN software pricing scheme.

LAN Media

LAN media refers to the type of wiring used for the physical connections. Most LAN systems use one of three general categories of wiring: coaxial cable, fiber optics, or twisted pair wires.

Coaxial cable systems use the same cables and connectors used by the cable television industry. The quantities of cables and connectors produced for the larger cable TV market has made the cost of these connections quite low. Coaxial cable-based systems are sold by IBM, 3Com, Quadram, and many other vendors.

Fiber optic media systems are rare but have a very practical advantage over coaxial systems. These systems are immune to interference caused by electrical or electronic devices. If a network will be operating in an environment where large machines or motors are turning on and off, or where there is strong radio frequency interference, then a network using fiber optics should be considered. Honeywell is one vendor with a mature fiber optics-based network. In many cases, a fiber optic cable can replace a coaxial cable without having to change network software.

The most familiar application of twisted pair wiring is the wiring between a desk telephone and the telephone wall unit. Inside of the installation jacket are four or more wires capable of carrying computer communications signals. This type of wire is low-cost and easy to install. The only limitations are capacity, speed, and susceptibility to interference from electronic devices.

Cost is the primary reason for considering a LAN's media. Capacity will only play a part in the decision process if a very fast data transfer is required or if future plans to add other activities, like video, to the network are needed. Costs for the three general types range from fiber optics, where small installations cost $10 per foot and $50 per termination; through coaxial cable, where installations cost $4 per foot and $40 per termination; to twisted pair, which costs pennies per foot and $2 per termination. If telephone wires are already in laboratory walls, then these wires can be used for a LAN.

In coaxial cable systems, the signals can be sent differently. Broadband systems can use the cable line as a radio link and transmit a wide number of frequencies over the cable. Each network station can then tune into the right frequency to receive messages just like a radio or television set tuner. Base-band systems use the cable line like a single wire and transmit data that is received at all network stations. Broad-band systems require more electronics, but the cables can send, in addition to data, television signals and other information needed to communicate. Base-band systems can send only digital data.

LAN Topology

LAN topology describes how information flows in the LAN. The topology of a network should be invisible to each user. Classical topologies are the bus, ring, and star configurations. The bus configuration uses a bus structure, in which each user is connected in line to the network. The ring configuration is similar to a bus except the "line" ends where it begins, thus ensuring that a transmission has been sent. The star configuration has a file server as its center, and information is passed along the network to various users located on the points of the star.

Current LANs for PCs use two different strategies. One strategy uses the internal disk drives contained in each PC. A second strategy uses a high-capacity (usually greater than 50 megabytes) external disk, or file server. A *file server* is a central computer that provides the total data storage for the entire LAN. Both methods require each PC on the network to have a special-function board inserted in one of the expansion slots. Each PC must also have software for network interaction. The internal disk drive method has lower initial start-up costs because the file server does not have to be purchased. But with the internal method, each PC on the network must contain much more software for network interaction, and data access is less uniform because various disks must be accessed, rather than one file server.

LANs allow multiple PCs to communicate freely with each other. Any PC on a LAN with the proper hardware and software can be used to send and receive data with a mainframe. In a LAN, these special PCs are called *gateways*. The hardware and software for the gateways are normally offered by the same company that produces the LAN. Through gateways, PCs on a LAN can communicate their data to the mainframe or receive data from a mainframe. LAN gateways have also been developed for the DEC–VAX family of minicomputers.

Future of LANs in Laboratories

A fully automated laboratory requires communication between each instrument. LANs promise to provide this communications link at a reasonable price. As more analytical instrument companies support the ability to send their data to a PC or provide the ability to send data and to be controlled via a PC, the full automation of laboratories will become possible. As new generations of analytical instruments are developed, the ability to communicate with PCs will be extremely important. Without the cooperation of the instrument manufacturers, the fully automated laboratory will never be attainable.

Products Mentioned in This Chapter

Informatics General Corporation, P.O. Box 1452, Canoga Park, CA 91304. (818) 716–1616

Kermit Distribution, Columbia University Center for Computing Activities, 7th floor, Watson Laboratory, 612 W. 15th St., New York, NY 10025.

IRMA Board, Digital Communications Associates Inc. (DCA), 1000 Alderman Dr., Alpharetta, GA 30201. (404) 442–4000

ZT–1444 GPIB Board, Ziatech Corporation, 3433 Roberto Court, San Luis Obispo, CA 93401. (805) 541–0488

Intelligent Interface, Nelson Analytical Inc., 10061 Bubb Rd., Cupertino, CA 95014. (408) 725–1107

LAN Vendors

IBM, P.O. Box 1328, Boca Raton, FL 33432. (800) 426–2468

3COM, 1365 Shorebird Way, P.O. Box 7390, Mountain View, CA 94039. (415) 961–9602

AST Research, 2121 Alton, Irvine, CA 92714. (714) 863–1333

Microsoft Networks, Microsoft, 10700 Northup Way, P.O. Box 97200, Bellevue, WA 98004. (206) 828–8080

Novell Inc., 1170 N. Industrial Park Dr., Orem, UT 84057. (800) 526–5463

Additional Reading

Campbell, Joe *Mastering Serial Communications;* SYBEX: Berkeley, CA, 1984.

Seyer, Martin D. *RS–232 Made Easy;* Prentice–Hall: Englewood Cliffs, NJ, 1984.

Chapter

On-Line Electronic Data Bases

powerful array of on-line data bases is available via modem and telephone links for the laboratory professional. More than 2500 information data bases are available to PC users. These data bases contain everything from the buying patterns of people ages 12–19 to the pedigrees, breeding, and racing records of all throroughbred race horses in North America since 1922. Data bases contain information for specific industries, such as *CAS ONLINE,* a service of the ACS that provides research abstracts from many chemical journals. General information services such as *CompuServe, The Source, Dow Jones News/Retrieval Service, Newsnet, Dialog,* and *Bibliographic Retrieval Services (BRS)* also provide information. The availability of this information in easy-to-retrieve form is another byproduct of the PC revolution. This type of information was once available only to organizations that could afford large research staffs and libraries. This situation meant that only people in the largest corporations or universities could have access to this information, but now it is available to anyone with a PC, modem, telephone, and communications software.

Using the same equipment and software, users can also communicate with other PC users. This communication can take the form of interacting one-on-one with another PC user via *CompuServe,* "listening" to an on-line interview, or reading bulletins on a local or distant bulletin board.

Getting Started

To get started, you will need a plug-in circuit board with an asynchronous communications port, a modem, a cable to connect

1000–4/87/0223$06.00/1   © 1987 American Chemical Society

the port to the modem, and communications software (Figure 10.1). Your PC may already have one or more communications ports because these ports are bundled on memory-expansion cards. Through PC–DOS, a PC can communicate with only two communications ports. If both communications ports are already in use to operate an X–Y plotter and communicate with an instrument, for example, then the port's use can be expanded by purchasing one or more switching boxes. These handy "black boxes" allow the user to connect two or more devices to the same port and switch from one device to the other (Figure 10.2). Switching can be done manually with a switch or electronically with the proper software. These handy devices are available from a number of vendors; the premier company is Black Box Corporation.

A modem that plugs directly into an expansion slot is also available. This feature is advantageous when little bench space is available, but the modem also uses an additional slot in the PC, does not allow the modem to be used with other computers in the lab, and adds additional heat to the computer. Modems also use one of the two possible communications lines (usually COM2:) the operating system allows. I suggest a Hayes or Hayes-compatible external modem that transmits and receives data at 1200 baud. These modems can also communicate at the slower 300-baud rate. The Hayes modem

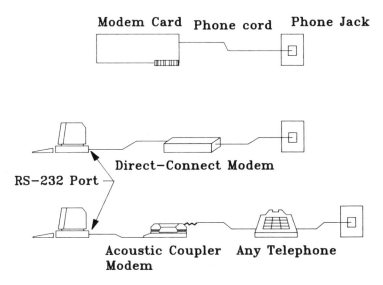

Figure 10.1 Components necessary for communicating with other computers and on-line data bases.

Figure 10.2   Using more than two devices from a communications port is made possible by using a switching box. Two- and four-way switching boxes are common. These boxes can also be used to share devices such as printers and plotters with others in a lab.

commands are a de facto standard that most communications software packages can use. Thus, the user need not write any programs for control of the features of a modem.

The communications software needed for modems is available from many sources. Two of the most popular are CrossTalk and PC–Talk III. The first versions of PC–Talk were the programs that pioneered the freeware and shareware method of distribution. The program's author, Andrew Fluegelman, encourages people to copy and disseminate PC–Talk III freely and send a contribution if his software is used. Thus, PC–Talk III is normally available from a users group public domain software library or from Headlands Press. PC–Talk is available from the Lab PC Users Group (see address at the end of this chapter).

Once this hardware and software are obtained, the user is ready to start communicating. Some of the available services are discussed in this chapter.

General Information and News

CompuServ

CompuServ, with almost 250,000 subscribers, is based in Columbus, OH, and is a wholly owned subsidiary of H&R Block. The suggested retail price to start the service is $39.95, which includes $25 of free time on the system. After that time is spent, the user must pay from $6 to $25 per hour depending on the time of day, baud rate, and means of access. Normally, a local telephone number provides access to the system, though in rural areas a long-distance call or access via Telenet or Tymnet may be needed. Telenet, owned by GTE Telenet in Reston, VA, and Tymnet, owned by Tymshare in Cupertino, CA, provide a network of additional local telephone lines. These services, also used by people with terminals to access their remote computers, cost an additional $2–$10 per hour but could be more economical if long-distance charges can be saved. *CompuServ* has no minimum monthly charge.

CompuServ provides an excellent introduction to the communication and information features of on-line services. The system provides electronic mailboxes, so the user can send other *CompuServ* users messages, letters, and reports. Many corporations use this low-cost service to communicate with other divisions or with their customers. *CompuServ* information service includes news from the Associated Press, *USA Today*, *The Washington Post*, and *The St. Louis Post–Dispatch*. Commodity and stock quotes are also included.

Many major software companies, such as Borland International and Lotus Development, provide on-line technical support and special interest group (SIG) information via *CompuServ*. Borland International has a 24-h SIG, and Lotus has the *World of Lotus*, which includes a category for scientific applications. These on-line services provide tips, examples, and answers to questions about products made by these companies.

CompuServ organizes information in named or numbered pages. You can access the desired pages by moving through menus to reach the destination or by using the FIND command if the name or number of the information wanted is not known. Once you know the page number or name by which the information can be accessed, type **go** and the target page name or number at any prompt marked with an exclamation point.

Considerable time and money can be saved by knowing what information is available, the page name, or the page number. The user does not have to move through menus to get to the desired information. To get a listing of this information, first type **go cis-9**, or select user information from the mail menu and choose the change terminal setting option that appears on the subsequent menu. Now, the user can specify how *CompuServ* will deliver its information to the PC screen. The important parameter to specify is a page size of 0 lines. If the user does not do this, *CompuServ* will pause and prompt the user to press return or the S key (for Scroll) whenever a certain number of lines has been displayed. With a specification of 0 lines, the information will flow continuously whenever the S key is pressed.

Now, type **go ind** to get to the *CompuServ* index. This index list can be saved on disk for printing after the information retrieval session. Set the communications software to capture incoming information, then select to list all indexed topics. After the first 11 items have appeared, you will be prompted to input a number or press the Return key for more choices. Do not do this. Instead, press the S key to make the index scroll continuously. If you follow the prompts and press Return, only the next 11 items will be read and displayed. To go immediately to a specific subject page, enter **go** and the page number or name in the right column.

The Source

The Source is in direct competition with *CompuServ* and provides many of the same services. *The Source* is a subsidiary of the Reader's Digest Association in conjunction with Control Data Corporation. The type and quantity of services available in the two major services are almost identical. The only difference is that *The Source* has a better introductory manual, thus the system is easier to use. *The Source* has about 60,000 users, many fewer than *CompuServ*. Thus, users are less likely to encounter delays during peak use periods.

The Source has one of the best electronic mail systems in the industry. The on-line conferencing system, Parti (short for participation), is one of the most popular services. The conferences are on-line and usually led by a noted expert in the field; they cover topics on computer applications, medicine, health, and other sciences.

Dow Jones News/Retrieval

The *Dow Jones News/Retrieval*, with more than 200,000 subscribers, is the leader in providing financial and corporate information. The service offers many corporate stock information data bases and provides real-time quotes from all major stock exchanges. (Real-time stock quotes are now availabe on *The Source* and *CompuServ* as well). Almost anything about corporate America, if it was published in the *Wall Street Journal, Barrons,* or in the annual report of one of 10,000 publicly held companies, is available from this service.

Information Data Bases

The two major pure information data base purveyors are *Dialog* and *Bibliographic Retrieval Services (BRS).* Both systems give users access to hundreds of independently produced data bases. Both services provide such diverse data bases as *Books in Print* (authors, subjects, and titles), *Magazine ASAP* (full text of *Time, Life, Sports Illustrated, People,* and more than 70 other publications), *NEWSEARCH* (drawn daily from more than 1700 newspapers, magazines, and periodicals), *INSPEC* (physics, electronic technology, and computer publications), and *CA Search* (the principal data base from Chemical Abstracts Service).

Most of the information on these systems is contained in huge data bases requiring very powerful and complex searching software. At user prices of $25–$300 per hour, the user can run up a bill of more than $100 without obtaining much information. The best way to approach these systems is just like any powerful software package. Before logging on for the first time, read the manuals carefully and possibly take a training course. Information and maintenance of the data bases is provided by companies called information providers (IPs). Virtually all IPs publish manuals and other materials to support the use of their data bases.

Dialog. *Dialog* is a subsidiary of Lockheed Corporation. In the early 1960s, Lockheed was a major contractor for NASA and needed to organize and retrieve the huge volume of information generated by the space program. *Dialog* was the result and now offers more than 200 data bases. Many of the *Dialog* data bases are available at a flat rate of $24 per hour during evenings and weekends through the *Knowledge Index.*

Bibliographic Retrieval Services. *BRS* is particularly strong in the medical, physical, and social sciences. *BRS* offers *BRS After Dark* and *Brkthru.* Both data bases offer much lower costs for evening and weekend use than the other services described.

Chemical Information Sources. *CAS ONLINE* is an on-line version of *Chemical Abstracts.* Data can be retrieved from the millions of journal articles, patents, and other chemical documents from national and international research. Searching can be performed using molecular structure, structural fragments, chemical names, molecular formulas, or CAS registry numbers.

Bulletin Boards

Most computer users groups have bulletin boards that can be accessed by phone, where users can get tips, programs, and information on how to use PCs better. Many groups have SIGs of users who apply the scientific applications of PCs. The lab PC users group has a bulletin board that can be accessed by members of the group. The bulletin board contains programs, information, and messages from group members. Questions about PC use in the lab can be answered by leaving a message. The group also has an extensive library of freeware, including the software mentioned in this book.

Products Mentioned in This Chapter

CrossTalk ($195), Microstuf Inc., 1000 Holcom Woods Parkway, Suite 440, Roswell, GA 30076. (404) 998–3998

PC–Talk III ($35), The Headlands Press, P.O. Box 862, Tiburon, CA 94920. Available through Lab PC Users Group

CompuServ ($39.95 for starter kit), CompuServ Information Service, 5000 Arlington Center Blvd., P.O. Box 20212, Columbus, OH 43220. (800) 848–8199; in Ohio (614) 457–0802

The Source ($0.43 per minute from 7 a.m. to 6 p.m. weekdays, and $0.18 per minute at all other times using 1200 baud; $10 per month minimum, $49.95 starter kit), The Source Telecomputing, 1616 Anderson Rd., McLean, VA 22102. (800) 336–3366 or (703) 734–7500

Dow Jones News/Retrieval ($75 initial subscription; $12 annual service fee; $0.18 per minute at 1200 baud), Dow Jones and Company, P.O. Box 300, Princeton, NJ 08540. (800) 257–5114

Bibliographic Retrieval Service (BRS) ($50 annual fee; $35 per hour on-line at 1200 baud), BRS Information Technologies, 1200 Rte. 7, Latham, NY 12110. (800) 345–4227

Dialog ($75 per hour on-line at 1200 baud, $50 for manual), Dialog Information Services, Inc., 3460 Hillview Ave., Palo Alto, CA 94304. (800) 334–2564

CAS ONLINE, Chemical Abstracts Service, Marketing Dept. 33685, P.O. Box 3012, Columbus, OH 43210. (800) 848–6538 or (614) 421–3600

Lab PC Users Group, 5989 Vista Loop, San Jose, CA 95124. (408) 723–0943

Black Box Corporation, P.O. Box 12800, Pittsburgh, PA 15241. (412) 746–5530

Chapter ELEVEN

Interfacing with the Real World

or acquiring data from an experiment or controlling a process, few standards are available, and thus much more information is needed about a particular application to select the proper interface. Selecting an interface for data acquisition and control requires the ability to read product literature and understand the terms used to describe these products. Product literature and advertising are filled with short phrases and acronyms describing the interfaces such as

- 6 analog output channels with 12-bit resolution,
- 24 bits of parallel TTL/CMOS-compatible digital I/O,
- A to D with 50 kHz throughput, and
- 8-channel differential or 16-channel single-ended input.

Selecting the proper interface requires knowledge about the application, the available hardware, and the available software. If you are new to interfacing, try to consult someone with experience. Reading every book and article written on the subject does not come close to the knowlege gained by experience. Computer interfacing for laboratory experimentation is still an art; the best references are the "artists" themselves. The time spent tracking someone down will produce huge dividends in the future. The vendors of interface cards are some of the best sources for the names of people who have successfully interfaced instruments or sensors to their PCs.

Before purchasing interfacing devices, be sure to see the products in action or have a demonstration of the exact configuration desired. Most PC interface card vendors have money-back guarantees if the board or software is returned within 30 days. Because of the low cost of these products, having them demonstrated in a laboratory is

1000–4/87/0231$06.00/1   © 1987 American Chemical Society

difficult. Expect to do some shopping by contacting another user or by viewing demonstrations at a seminar or trade show.

Interface Basics

Interfaces either transfer data in the same form from one computer to another or convert the data from its present form into another form so data can more easily be communicated. Transferring data can be performed by many types of physical media. The most common media are wires (telephone) and air (radio waves, television transmissions, and microwaves). Normally, data transfer is performed efficiently with little user interaction. Interfaces that perform data conversion require the user's attention because users are an integral part of performing any application. Data conversion interfaces can be hardware, software, or a combination of both hardware and software. Most PC interfaces are a combination of both hardware and software; thus, a number of programmable data conversions can be performed. Choosing and correctly using the proper type of converting hardware and software are the key ingredients in successful data acquisition and subsequent data handling on a PC.

Data Conversion

Most of the world around us is perceived as continuous or analog. Representing physical phenomena like the position of an object, amount of light, pressure, temperature, or force in a digital PC requires conversion of the data (Figure 11.1).

The first step in the most common method of conversion of a physical phenomenon into computer-readable form is to convert the physical stimulus into an electrical signal. The electrical signal can then be captured, manipulated, and displayed. Of the many characteristics of electrical signals, three groups are used most for data conversion: amplitude (A), time interval (dT), and digital (D). The amplitude group includes all electrical quantities such as voltage, resistance, capacitance, and current. In all these examples, the magnitude of the quantity is related to the measured phenomenon. The time interval category includes those electrical signals whose time relationship between parts of the electrical waveform, or between waveforms, is a measure of the physical phenomenon. These include the pulse width, frequency, or phase angle of the signal. Both the amplitude group and the time interval group are continuous functions of the measured phenomenon. A *digital signal* is a specific

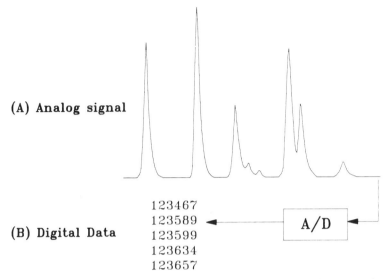

Figure 11.1 Data conversion. (A) Continuous analog signal (continuous range and domain) and (B) converted digital data (discrete range and domain).

number that discontinuously represents the measured phenomenon. A PC can easily acquire and manipulate this type of signal. Producing a digital signal requires that the other data types undergo conversion. All communication between the real world and a PC can be described as interconversions between these groups, or *data domains* (Figure 11.2).

Interdomain Conversion

Viewing communication with a PC as *interdomain conversions* enables the user to understand the function of components necessary for a particular application. For example, a temperature can be communicated to a PC. Figure 11.3 shows the components of a typical thermistor, a thermoresistive temperature sensor, interfaced to a PC. The first step is a physical phenomenon-to-electrical amplitude (P–A) conversion from temperature into an electrical signal in which the amplitude of the voltage produced by a Wheatstone bridge circuit is measured. The voltage is then converted into a frequency (A–dT) that is counted by a digital frequency meter (dT–D), and the digital value is sent to the IBM PC's memory (D–D).

Sending data from one computer to another computer via telephone lines and modems also requires a number of interdomain

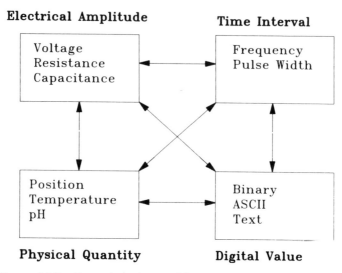

Figure 11.2   Data domains and interconversions required for digital data acquisition.

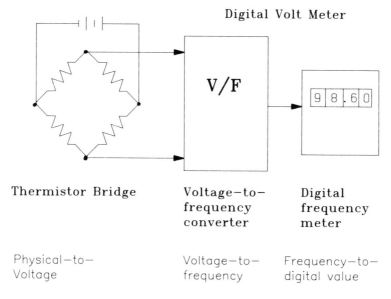

Figure 11.3   Digital temperature measurement using a thermistor transducer, voltage-to-frequency converter, and digital frequency meter.

conversions. Figure 11.4 shows this process. First, the digital data is converted into a transmittable code such as ASCII or EBCDIC by a D–D conversion. This code is converted into electrical frequency by the transmitting modem (D–dT). The electrical frequencies are transported over telephone lines or microwaves to the receiving modem, where they are converted back to digital data by the receiving modem (dT–D). The digital code is then converted into the digital representation for the receiving computer (D–D).

Interfacing Considerations

Although computers and computer interfaces may solve problems, they can just as easily inject errors. These errors take two forms: *transfer errors* and *data conversion errors*. Transfer errors can normally be detected and the data retransmitted and captured. Good computer interface software and hardware design will include error trapping and proper error recovery. The same techniques that are used in performing file transfers from one computer to another, such as the XMODEM or Kermit protocols, should be implemented so transfer errors do not occur. Of course, sometimes this approach is not possible without additional work, but the integrity of data is at stake, and all errors should be omitted if possible.

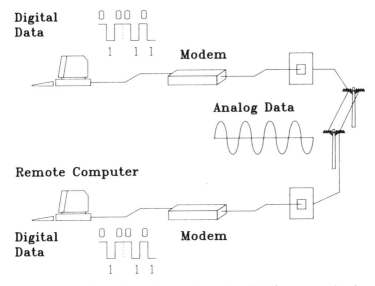

Figure 11.4   Interdomain transfers for digital communication over telephone lines.

Data conversion errors are normally more subtle and are actually designed into the interface. Data conversion processes in communication are notorious for producing and injecting errors in a computing process. At almost every data conversion point in an interface, the data conversion may be performed inaccurately. Some common data conversion devices and considerations when selecting a device will now be discussed.

Transducers

Transducers convert a physical stimulus (position, light, temperature, acceleration, and force) into an electrical signal. One or more of the electrical signal's measurable characteristics (voltage, resistance, current, and capacitance) are proportional to the value of the measured stimulus. How precisely a transducer performs data conversion is described by its precision, linearity, frequency response, and hysteresis.

The *precision* of the transducer measures the reproducibility of a measurement. If an identical stimulus is repeatedly applied to a transducer, an identical result should be produced. Precision is a measure of how well a given transducer meets this ideal condition. Precision is normally reported as a standard deviation or percent standard deviation in converted units.

Linearity is a measure of how well a transducer meets an ideal calibration line or predictable function. A calibration line, or curve, is constructed by plotting stimulus values versus the transducer response. An ideal transducer will always respond to a stimulus with the corresponding values on this line, or curve. Real transducers have good linearity over only a specific range of stimulus values.

Frequency response describes how quickly the transducer can track changes in the applied stimulus.

Hysteresis is the difference in the readings a transducer reports for the same stimulus depending on whether the stimulus value is approached from below (ascending) or from above (descending).

All of these parameters are described for a component in a specific temperature range. If higher or lower temperatures are experienced by the device, the instrument can easily deviate significantly from the specifications.

Analog Voltage-to-Digital Converter

Hundreds of analog voltage-to-digital data (A/D) converters are available for the IBM PC. The basic application environment for one of these devices is diagramed in Figure 11.5. A stimulus (pressure, temperature, position, or light) is applied to a transducer. The transducer produces an output voltage E1. This voltage is then amplified to become E2, which is compatible with the input voltage of the A/D converter. The analog voltage is then converted into a digital value. When an analog voltage is converted to a digital value, two continuous data domains, voltage and time, are digitized. How precisely an A/D can perform this data conversion is described by the same terms mentioned for transducers plus sampling rate, resolution, and throughput.

Sampling rate is the number of A/D conversions performed per second. When an analog signal is converted (Figure 11.6), error is inherent in the new representation. This error can be reduced to a negligible level if a sufficiently large number of samples are taken. Faster sampling is not always a practical solution to minimize this error.

A good rule of thumb for setting the sampling rate, Nyquist's criterion, is to sample the signal at twice the desired rate. In cases where it is difficult to select this rate, different sampling rates must be tried. The difference that this rate has on the stored data should then be observed. Figure 11.7 shows the stored data obtained by sampling

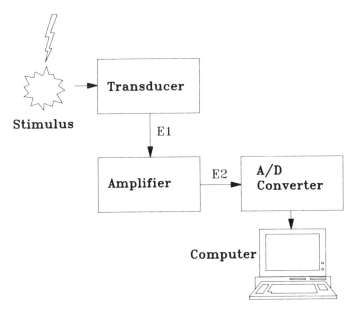

Figure 11.5   Typical digital instrumentation system.

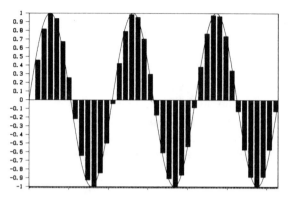

Figure 11.6   Errors caused by data conversion are represented by the difference from the baseline.

the same sine wave at slow and fast rates. The importance of the correct sampling rate on this data can be seen clearly.

Resolution, or *dynamic range,* dictates how many discrete values the analog level can be assigned. This value is normally reported as the number of bits the resulting digital value can be assigned. The most comon are 8-, 12-, and 16-bit A/D converters. If voltage–frequency-type A/Ds are used, the resolution is the number of digits in the digital counter. The more bits of resolution, or digits in the counter, the more precise the digital reading.

A bit can have a value of either 0 or 1. A single bit can represent only two (2^1) different values. By combining more bits, more representations can be described. For example, 4 bits can represent 2^4, or 16 input levels; 8 bits can represent 256 levels; 12 bits 4,096; and 16 bits can describe 2^{16}, or 65,536 different voltage levels. The examples in Figures 11.8–11.11 show how important this aspect of data conversion is to the correct interpretation of experimental results. The figures show a series of peaks, or *events,* that become smaller in amplitude. The largest amplitude event is recorded by all of the data conversions. However, the following events become more difficult to perceive except in the highest resolution conversion.

Another way to view the same error in conversion is to calculate the minimum measured difference that can be recorded in the analog signal. On a 5-V signal using a 4-bit A/D, the minimum measured difference that can be digitally recorded is 5 V/2^4, or 5 V/16 = 312 mV. Similarly, for 8-, 12-, and 16-bit A/Ds, the minimum measured difference would be 5 V/2^8, or 5 V/256 = 19.53 mV; 5 V/2^{12}, or

105 Points 0 to 2 PI
315 points total

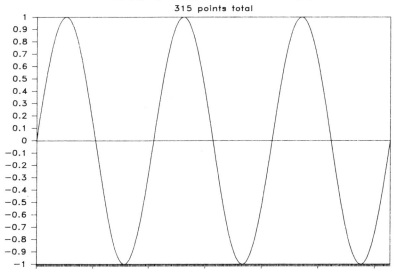

52 points from 0 to 2 PI
156 points total

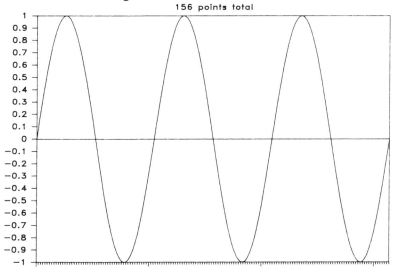

Figure 11.7   Effects of using different sampling on the same sine curve. Sampling rates are halved for each graph in this series. (Continued on next page.)

26 points from 0 to 2 PI
78 points total

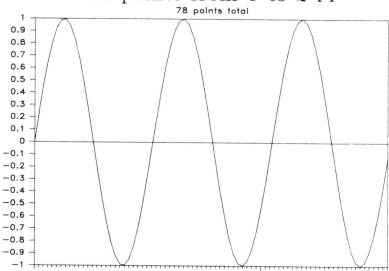

13 points from 0 to 2 PI
39 points total

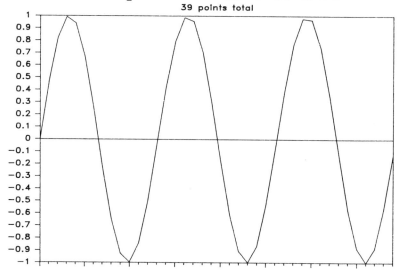

Figure 11.7—Continued.

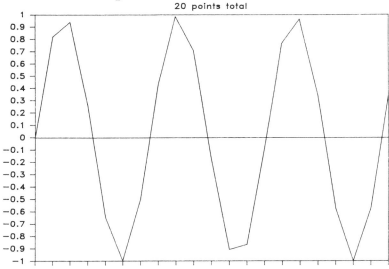

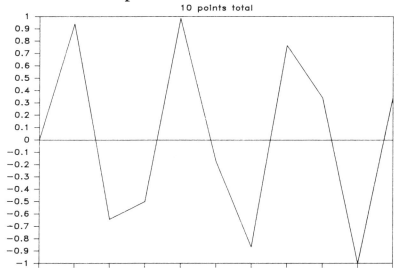

Figure 11.7—Continued.

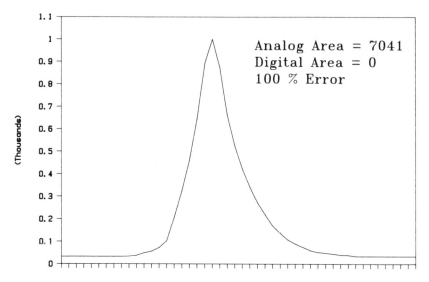

Figure 11.8   Digital result of an analog signal with an 8-bit A/D converter.

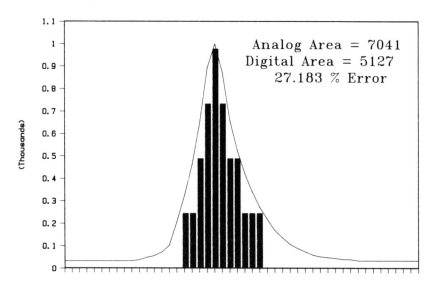

Figure 11.9   Digital result of an analog signal with a 12-bit A/D converter.

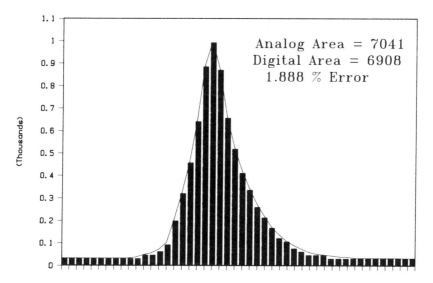

Figure 11.10   Digital result of an analog signal with a 16-bit A/D converter.

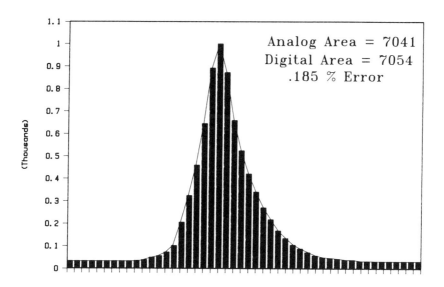

Figure 11.11   Digital result of an analog signal with a 20-bit A/D converter.

5 V/4096 = 1.22 mV; and 5 V/2^{16}, or 5 V/65,536 = 0.076 mV, respectively. All events in which the voltage change is below the minimum measured difference would go undetected.

For a specific application, A/D sampling rate and resolution must be sufficient to detect and capture the physical event observed. For example, the sampling rate and resolution necessary to monitor the changes in temperature in a greenhouse over 24 hours are quite different from those necessary to measure the temperature changes that occur in a 1-second interval inside a cylinder of a turbocharged engine. Consider increasing resolution if the computing methods used on the stored data will be squaring or taking the square roots of the resulting values, for example, when fitting a least-squares line or curve. Performing these conversions on already approximate values limited by the hardware's resolution can produce large unpredictable swings in the resulting data.

Throughput, or acquisition rate, is the amount of data collected during a specific amount of time. Most A/Ds report this specification in samples per second. Be aware that the values reported normally represent the maximum speed that the hardware on the interface card can perform the conversions. This maximum speed is the upper limit of the acquisition rate. These specifications rarely consider the additional time per sample required for multiplexer settling, also called sample and hold times. In addition, the time to display the data on a monitor or store the data on disk must be factored to obtain the true time required per sample. The true throughput, or acquisition rate, is the total time per sample required to acquire, display, and store data. On a conventional IBM PC running at a 5-MHz clock rate, a realistic throughput using conventional techniques supplied by interface vendors is about 500 samples per second. This acquisition rate can be increased substantially by using the faster PC/AT or a PC/AT clone; by using or writing faster software, or if only a burst of data acquired over a short period of time is needed, by acquiring data directly into memory and then storing the data after the acquisition period is over.

Analog-to-Digital Conversion of Chromatographic Data

The analog signal from a chromatographic detector is probably the most monitored analog signal in most chemical laboratories. This signal provides an excellent example for describing what should be considered when choosing an interface for a PC to the real world.

Chromatography has a wide range of applications, from a demanding analysis of a sample with hundreds of distinct components to an analysis with just one or two compounds. Chromatographic applications depend on the sample analyzed. Some chromatography experiments are performed for just qualitative results; others are performed for both qualitative and quantitative results. Similarly, the A/D conversion hardware and software selected should match the application. If all types of chromatographic data analysis will be performed, then the A/D hardware and software obtained must perform at all levels.

Chromatographic data acquisition on demanding samples that require both qualitative and quantitative analysis requires two features that are normally not found in conventional A/D converters. First, these experiments require an A/D with a large dynamic range, or high resolution. This large dynamic range is required because in the same sample, chromatographic peaks can be very small and represent low concentration components; also, very large peaks can be found that represent high concentration components. Covering this wide concentration range and also obtaining accurate quantitative data requires an A/D with a dynamic range of 1 million, or 10^6, to cover the normal 0–1-V detector output range. This means the analog signal is converted into one of 1 million possible values. High resolution is also important for chromatographic data because most of the data processing needed to find the start, apex, and end of a peak requires data value comparison and curve fitting, both of which use the squares and square roots of the values. Detecting these peak parameters accurately requires data that has been collected by using a high-resolution A/D.

The second feature not usually found in conventional A/D converters is that the conversions represent the integral under the analog signal, usually called an *integrating A/D,* rather than single, discrete, converted samples. This requirement is also dictated by the application. A chromatographic detector measures the concentration of a component eluting from the chromatograph. The concentration of a component is proportional to the area under the peak in the chromatogram representing the compound. An integrating A/D measures the area under the signal by monitoring the signal almost continuously through each sample time. A discrete A/D will monitor the signal for only a small percentage of the sample time, usually about 20%, and will assume that the analog signal was constant during the balance of the sample time. This type of A/D is fine when analog signals, such as temperature or pressure monitors that do not change over the sampling time, are measured, but chromatographic data requires better accuracy.

Both of these application-specific requirements are fulfilled by using an A/D with a voltage-to-frequency (V/F) converter. The analog signal voltage is monitored, and the V/F converts the voltage into a frequency of square waves, or pulses. The higher the input voltage, the higher the frequency of *pulses*. The pulses are counted in a specific sampling time period, and when the time period has elapsed, the counter's value is read into memory for storage, the counter is set to zero, and a new sampling period is started. This regime is continued for each data sample. This method of A/D conversion meets both of the criteria for a demanding chromatographic experiment. The analog signal is almost continuously monitored, and whereas the counter is read and set to zero at the end of each sampling period, the analog signal is not monitored. If the counter contains enough digits, six for the chromatographic application to obtain a dynamic range of 10^6, the necessary resolution will be attained.

If the chromatographic application is not as demanding, A/D converters with less resolution and dynamic range can be used. If you are not interested in quantitation, you may even consider using a nonintegrating A/D. Evaluate the application needs and use the proper data acquisition and interfacing devices. Data acquisition is a critical step in all experimentation because poorly collected data almost always leads to poor results and false deductions, no matter how sophisticated the data analysis routines are.

Digital-to-Analog Converters

These converters output an analog voltage for a specific digital value. All of the characteristics described for evaluating transducers and A/Ds are used to rate D/As.

The analog signals generated by a D/A are normally used to control a device. The voltage level sent to the device can cause the device to operate at a specific rate. For example, many HPLC pumps are controlled by an analog signal. The flow rate provided is determined by the voltage sent to the pump. To change the flow rate, simply send a different analog signal level.

Digital input/output, also called digital I/O or DIO, systems allow the user to accept or supply parallel digital data. Parallel data requires a separate connection for each bit. The most common use for this type of interface is for sending and receiving binary coded decimal (BCD), though each individual line (connection) can be used, for example, to switch on or off a device.

Data Acquisition Software

Most of the interface manufacturers provide some software for simple data acquisition. Normally, these programs are enough to get someone started and ensure that the hardware is operating correctly. For creating complete systems including data acquisition, data display, processing, and storage, three software packages should be investigated. The packages are Labtech Notebook, ASYST, and ASYSTANT.

Labtech Notebook

Labtech Notebook was developed by Laboratory Technologies. Labtech Notebook is primarily a data acquisition system that acquires data from a number of interface cards, RS–232C, and GPIB. The data is then stored on disk for further processing within Labtech Notebook or by other programs such as Lotus 1–2–3 or a custom program. Labtech uses menus similar to the popular Lotus 1–2–3 menu selection system to set up data acquisition. Labtech Notebook supports data acquisition rates as high as 600 samples per second with data displayed on the PC graphics monitor; it can acquire data as quickly as 1000 samples per second in high-speed mode with some restrictions (Figure 11.12).

ASYST and ASYSTANT

ASYST and ASYSTANT were developed by Adaptable Laboratory Software and are distributed and supported by Macmillan Software. ASYST is an interactive programming environment with built-in functions for common data acquisition, data display, and data reduction activities. The user can then assemble these building blocks into a custom application. Data can be acquired from a number of interface cards through RS–232C and GPIB. Data can be displayed graphically by using various graphing functions that include three-dimensional plots. Data analysis built-in functions include fast Fourier transform, data smoothing (Figure 11.13), integration, differentiation, curve fitting, statistical analysis, differential equation solving, and matrix operations.

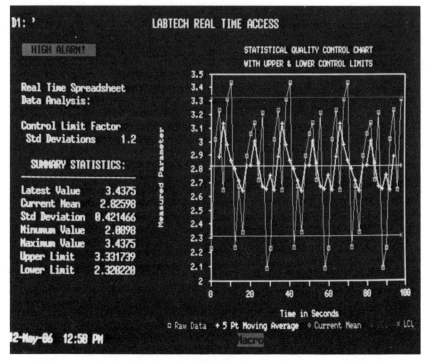

Figure 11.12   Labtech Notebook's real-time access. Graphs can be generated as data is obtained. (Courtesy of Laboratory Technologies.)

ASYSTANT is menu-driven software with built-in functions similar to those in ASYST. By making menu selections, the user can acquire, display, and analyze data.

Future of the PC in the Chemical Laboratory

This book has introduced the major applications that a PC can perform in a laboratory or office. In the future, the number of PC applications should increase dramatically, and the potential for personal computing in the laboratory will be great. The major trend driving this increase in the number of applications is the continued increase of computing power at lower cost. Applications that could be performed only by mini- or mainframe computers a few years ago can now be performed by PCs.

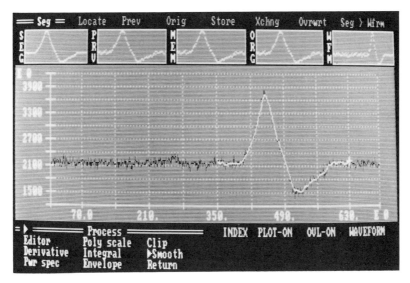

Figure 11.13 ASYST data analysis functions include data smoothing. (Courtesy of Macmillan Software.)

Lower cost for computing power leads to applications never before performed on any level of computer. Applications like desk-top publishing are now a reality and provide better ways for chemists to communicate. The ability to mix text and graphics, a need for all chemists, is now available at very attractive prices. The important ingredient for all of these PC-based solutions is brainware. The user must supply the brainware necessary to use these powerful tools. Our challenge in the future is not just to use these powerful tools, but always to look for the next application. The PC has provided us with fantastic opportunities for personal growth and control over our world. Let us hope this power can be used efficiently for constructive applications that can make all our lives better in the future.

Products Mentioned in This Chapter

LABTECH NOTEBOOK ($895), Laboratory Technologies Corp., 255 Ballardvale St., Wilmington, MA 01887. (617) 657–5400

ASYST (four modules, $2195), Macmillan Software Co., 866 Third Ave., New York, NY 10022. (800) 348–0033

ASYSTANT ($495) and ASYSTANT Plus ($895), Macmillan Software Co., 866 Third Ave., New York, NY 10022. (800) 348–0033

Selected Manufacturers of Data Acquisition and Control Interfaces for the IBM PC

MetraByte Corp., 254 Tosca Dr., Stoughton, MA 02072. (617) 344–1990

Data Translation, 100 Locke Dr., Marlboro, MA 01752. (617) 481–3700

Cyborg Corp., 55 Chapel St., Newton, MA 02158. (617) 964–9020 or (800) 343–4494

Strawberry Tree Computers, 1010 West Fremont Ave., Sunnyvale, CA 94087. (408) 736–3083

Keithley Data Acquisition and Control, 28775 Aurora Rd., Solon, OH 44139. (800) 552–1115

Additional Reading

Malmstadt, M. V.; Enke, C. G. *Digital Electronics for Scientists;* W.A. Benjamin: New York, 1969.

Dessy, R. E. *The Electronic Laboratory;* American Chemical Society: Washington, DC, 1985.

Leibson, S. "The Input/Output Primer", *BYTE,* February–July 1982.

Appendix

Cell by Cell Listings for Lotus 1-2-3 Examples

Area Percent

B1: **Liquid Chromatography Data**
B2: **Area Percent Report**
A3: \=
B3: \=
C3: \=
D3: \=
E3: \=
A4: **'Sample:**
B4: **'QA #2345**
D4: **'Date:**
E4: **(D1) @DATE(85,3,4)**
A5: **'Column:**
B5: **'Sephadex**
D5: **'Operator:**
E5: **'GIO**
A6: **'Solvent:**
B6: **'Methanol/Water**
D6: **'Inst No.:**
E6: **'LC #1234**
A7: **'Notes:**
B7: **'Sample vial seal broken**
    during shipment
A8: \=
B8: \=
C8: \=
D8: \=
E8: \=

A9: **'Peak No.**
B9: **'Ret. Time**
C9: **^Area**
D9: **'% Total Area**
A10: **1**
B10: **2.34**
C10: **12345**
D10: **(F3) 100*C10/$C$19**
A11: **2**
B11: **2.67**
C11: **34567**
D11: **(F3) 100*C11/$C$19**
A12: **3**
B12: **3.68**
C12: **78904**
D12: **(F3) 100*C12/$C$19**
A13: **4**
B13: **4.98**
C13: **5674**
D13: **(F3) 100*C13/$C$19**
A14: **5**
B14: **6.89**
C14: **12356**
D14: **(F3) 100*C14/$C$19**
A15: **6**
B15: **9.78**
C15: **896654**

D15: (F3) 100*C15/$C$19

A16: 7

B16: 10.76

C16: 9223

D16: (F3) 100*C16/$C$19

A17: 8

B17: 13.45

C17: 14568

D17: (F3) 100*C17/$C$19

C18: \-

D18: (F3) \-

C19: @SUM(C10..C17)

D19: (F3) @SUM(D10..D17)

Single External Standard Quantitation

B1: 'Liquid Chromatography Data

B2: 'Single Level External Standard

A3: \=

B3: \=

C3: \=

D3: \=

E3: \=

F3: \=

G3: \=

A4: 'Sample:

B4: 'QA #2345

D4: 'Date:

E4: (D1) @DATE(85,3,4)

A5: 'Column:

B5: 'Sephadex

D5: 'Operator:

E5: 'GIO

A6: 'Solvent:

B6: 'Methanol/Water

D6: 'Inst No.:

E6: 'LC #1234

A7: 'Notes:

B7: 'Sample vial seal broken
   during shipment

A8: \=

B8: \=

C8: \=

D8: \=

E8: \=

F8: \=

G8: \=

A9: 'Compound

B9: 'Ret. Time

C9: ^Area

D9: ^Amount

E9: 'Resp Factor

F9: ^Amt %

G9: 'RT Delta%

A10: 'Methane

B10: 2.34

C10: 12345

D10: (F3) +C10/E10

E10: +E24

F10: (P3) +D10/$D$19

G10: (P3) (B24-B10)/B24

A11: 'Ethane

B11: 2.67

C11: 34567

D11: (F3) +C11/E11

E11: +E25

F11: (P3) +D11/$D$19

G11: (P3) (B25-B11)/B25

A12: 'Propane

B12: 3.68

C12: 78904

D12: (F3) +C12/E12

E12: +E26

F12: (P3) +D12/$D$19

G12: (P3) (B26-B12)/B26

A13: 'Butane

B13: 4.98

C13: 5674

D13: (F3) +C13/E13

E13: +E27

F13: (P3) +D13/$D$19

G13: (P3) (B27-B13)/B27

A14: 'Pentane

B14: 6.89

C14: 12356

D14: (F3) +C14/E14

E14: +E28

F14: (P3) +D14/$D$19

G14: (P3) (B28-B14)/B28

A15: 'Hexane

B15: 9.78

C15: 896654

D15: **(F3)** **+C15/E15**
E15: **+E29**
F15: **(P3)** **+D15/$D$19**
G15: **(P3)** **(B29-B15)/B29**
A16: **'Heptane**
B16: **10.76**
C16: **9223**
D16: **(F3)** **+C16/E16**
E16: **+E30**
F16: **(P3)** **+D16/$D$19**
G16: **(P3)** **(B30-B16)/B30**
A17: **'Octane**

B17: **13.45**
C17: **14568**
D17: **(F3)** **+C17/E17**
E17: **+E31**
F17: **(P3)** **+D17/$D$19**
G17: **(P3)** **(B31-B17)/B31**
C18: **\-**
D18: **(F3)** **\-**
F18: **\-**
C19: **@SUM(C10..C17)**
D19: **(F3) @SUM(D10..D17)**
F19: **(P2) @SUM(F10..F17)**

Linear Least-Squares Fit

A1: **'Calibration Data**
   for 1,3 dichlorobenzene
A3: **"Concentration**
B3: **"Area**
A6: **100**
B6: **21000**
C6: **+A6*B6**
A7: **200**
B7: **30400**
C7: **+A7*B7**
A8: **300**
B8: **39700**
C8: **+A8*B8**
A9: **400**

B9: **48900**
C9: **+A9*B9**
A10: **500**
B10: **61000**
C10: **+A10*B10**
A13: **'Slope of Line:**
B13: **(@SUM(PRODUCT)/**
   @COUNT(CONC)-@AVG
   (CONC)*@AVG(AREA))/
   @STD(CONC)^2
A14: **'Y-Axis Intercept:**
B14: **@AVG(AREA)-B13***
   @AVG(CONC)

Instrument Quality Control Chart

A1: **'Data**
B1: **'Avg**
C1: **'Avg-1Std**
D1: **'Avg+1Std**
B2: **@AVG($DATA)**
C2: **+B2-@STD($DATA)**
D2: **+B2+@STD($DATA)**
A3: **100**
B3: **@AVG($DATA)**
C3: **+B3-@STD($DATA)**
D3: **+B3+@STD($DATA)**
I3: **6**
A4: **110**
B4: **@AVG($DATA)**
C4: **+B4-@STD($DATA)**
D4: **+B4+@STD($DATA)**

I4: **7**
A5: **120**
B5: **@AVG($DATA)**
C5: **+B5-@STD($DATA)**
D5: **+B5+@STD($DATA)**
I5: **8**
A6: **105**
B6: **@AVG($DATA)**
C6: **+B6-@STD($DATA)**
D6: **+B6+@STD($DATA)**
I6: **9**
A7: **130**
B7: **@AVG($DATA)**
C7: **+B7-@STD($DATA)**
D7: **+B7+@STD($DATA)**
I7: **10**

A8: 112
B8: @AVG($DATA)
C8: +B8-@STD($DATA)
D8: +B8+@STD($DATA)
I8: 11
A9: 111
B9: @AVG($DATA)
C9: +B9-@STD($DATA)
D9: +B9+@STD($DATA)
I9: 12
A10: 109
B10: @AVG($DATA)
C10: +B10-@STD($DATA)
D10: +B10+@STD($DATA)
I10: 13
A11: 108
B11: @AVG($DATA)
C11: +B11-@STD($DATA)
D11: +B11+@STD($DATA)
I11: 14
A12: 108
B12: @AVG($DATA)
C12: +B12-@STD($DATA)
D12: +B12+@STD($DATA)
I12: 15
A13: 110
B13: @AVG($DATA)
C13: +B13-@STD($DATA)
D13: +B13+@STD($DATA)
I13: 16
A14: 111
B14: @AVG($DATA)

C14: +B14-@STD($DATA)
D14: +B14+@STD($DATA)
I14: 17
A15: 109
B15: @AVG($DATA)
C15: +B15-@STD($DATA)
D15: +B15+@STD($DATA)
I15: 18
A16: 110
B16: @AVG($DATA)
C16: +B16-@STD($DATA)
D16: +B16+@STD($DATA)
I16: 19
A17: 109
B17: @AVG($DATA)
C17: +B17-@STD($DATA)
D17: +B17+@STD($DATA)
I17: 20
A18: 108
B18: @AVG($DATA)
C18: +B18-@STD($DATA)
D18: +B18+@STD($DATA)
I18: 21
B19: @AVG($DATA)
C19: +B19-@STD($DATA)
D19: +B19+@STD($DATA)
E19: 'Average
F19: 'Avg-1Std
G19: 'Avg+1Std
H19: 112
I19: 22

Lotus Macro

AA1: 1+1
AB1: 'Counter
AA2: 2
AB2: 'Comparison
AA3: '\P
AB3: "/RECOUNTER~
AB4: "/XNEnter Number
      of copies:~COMPARISON~

AB5: "/PPR{?}~Q
AB6: "/XICOUNTER>
      =COMPARISON~/XQ
AB7: "/PPAGPQ
AB8: "{GOTO}
      COUNTER~{EDIT}+1~
AB9: "/XGAB6~

ICP Data for Graphics Application

A1: 1474
B1: 1179

C1: 884
A2: 1732

B2: **1385**
C2: **1039**
A3: **2116**
B3: **1692**
C3: **1269**
A4: **2628**
B4: **2102**
C4: **1576**
A5: **3364**
B5: **2691**
C5: **2018**
A6: **4528**
B6: **3622**
C6: **2716**
A7: **6636**
B7: **5308**
C7: **3981**
A8: **10896**
B8: **8716**
C8: **6537**
A9: **19632**
B9: **15705**
C9: **11779**
A10: **36464**
B10: **29171**
C10: **21878**
A11: **75264**
B11: **60211**
C11: **45158**
A12: **152256**
B12: **121804**
C12: **91353**
A13: **257664**
B13: **206131**
C13: **154598**
A14: **333440**
B14: **266752**
C14: **200064**
A15: **353344**
B15: **282675**
C15: **212006**
A16: **300160**
B16: **240128**
C16: **180096**
A17: **204464**
B17: **163571**

C17: **122678**
A18: **114560**
B18: **91648**
C18: **68736**
A19: **58048**
B19: **46438**
C19: **34828**
A20: **33392**
B20: **26713**
C20: **20035**
A21: **22448**
B21: **17958**
C21: **13468**
A22: **16416**
B22: **13132**
C22: **9849**
A23: **13040**
B23: **10432**
C23: **7824**
A24: **10356**
B24: **8284**
C24: **6213**
A25: **8520**
B25: **6816**
C25: **5112**
A26: **7216**
B26: **5772**
C26: **4329**
A27: **6244**
B27: **4995**
C27: **3746**
A28: **5392**
B28: **4313**
C28: **3235**
A29: **4744**
B29: **3795**
C29: **2846**
A30: **4180**
B30: **3344**
C30: **2508**
A31: **3756**
B31: **3004**
C31: **2253**
A32: **3344**
B32: **2675**
C32: **2006**

Pie-Chart Data

A1: 'Design & Development
D1: 35
A2: 'Analysis & Simulation
D2: 15
A3: 'Test & Lab Automation

D3: 20
A4: 'Administration
D4: 10
A5: 'PC Applications
D5: 20

Stacked-Bar Chart Data

A1: 'Data
A3: 100
A4: 110
A5: 120
A6: 105
A7: 130
A8: 112
A9: 111
A10: 109
A11: 108
A12: 108
A13: 110
A14: 111
A15: 109
A16: 110
A17: 109
A18: 108
A20: 60
A21: 70
A22: 80
A23: 65
A24: 90
A25: 72
A26: 71
A27: 69

A28: 68
A29: 68
A30: 70
A31: 71
A32: 69
A33: 70
A34: 69
A35: 68
A40: 50
A41: 60
A42: 70
A43: 55
A44: 80
A45: 62
A46: 61
A47: 59
A48: 58
A49: 58
A50: 60
A51: 61
A52: 59
A53: 60
A54: 59
A55: 58

Glossary

active talker general-purpose interface bus (GPIB) device that can send messages to other devices

activity task that is needed to complete a project

A/D conversion conversion of an analog signal to a digital value

ad hoc query data-retrieval method that allows specific information to be obtained upon the user's request; found in data base management systems

address memory location in the central processing unit

analog signal signal with a continuous range and domain (e.g., electric voltage)

application environment application program that runs on PC–DOS and provides an advanced user interface

application software programs that perform specific applications such as word processing, spreadsheet analysis, and graphics

ASCII file a file in American Standard Code for Information Interchange, in which each symbol in the file is converted to a binary form of one byte, or eight bits

asynchronous communications port device that sends and receives data one bit at a time; also called serial port

attribute column in a data model table in a data base management system

baud rate the number of bits per second transmitted in a communication

bit a shortened form of "binary digit", which can have a state, or value, of zero or one

brainware knowledge and training in hardware and software use

257

bus   connection that links peripheral devices to a personal computer

byte   cluster of eight bits

cell   intersection of a column and a row on a spreadsheet

central processing unit (CPU)   device on a personal computer that performs the calculations and orchestrates where data is to be read from, stored, or sent

checksum   value sent by a computer after each block of data is transmitted to detect errors in data transmission

chip   device containing integrated circuits

communications processor   second computer designed to control communication between the host computer and the terminals

compiler   language that transforms source code into machine or object code

critical path   sequence of a project that takes the longest time

cursor   lighted box on the computer screen that shows the user where data can be entered or changed

data base   collection of information organized and presented to have a specific purpose

data base management system (DBMS)   programs that organize and manipulate data

data domain   environment that contains data; examples include personal computers and the real world

data log   list of simple analysis information in a data logbook program

data logbook program   program used to track the results of tests performed on final or intermediate manufactured products

data parsing   capability that allows an ASCII file containing labels and numbers to be imported

data query   section in a logbook that is used for summarizing and extracting data from the data log information

digital signal   specific number that discontinuously represents a measured phenomenon

disk drive   device that provides read–write storage for programs and data

drawing program   program that allows graphs and charts to be generated from data

driver   program that controls input and output devices such as monitors, keyboards, printers, plotters, and disk drives

duplex transmission of data in two directions at the same time

electronic mail capability of a network that allows users to communicate with each other

event peak in a digital representation of an analog signal

field category of information in a data base

file storage device where fields and records are kept on a disk; also called a relation

filing program type of data management program that allows data stored in a disk file to be created, edited, and reported

freeware programs that are distributed free of charge; if the program is used regularly, the user is asked to make a donation to support the program

frequency response measure of how quickly a transducer can track changes in the applied stimulus

gateway personal computer in a local area network that can be used to send and receive data from a mainframe or minicomputer

general-purpose interface bus (GPIB) device designed to provide mechanical, electrical, timing, and data compatibilities among all devices adhering to a standard

half duplex transmission of data in two directions at different times

handshaking communication of status information from one device to another

hardware physical equipment that makes up a computer

heavyweight data management program that performs like mainframe computer data base systems

hot key program program that is loaded into the memory and can be initiated with the touch of one key

hysteresis difference in the transducer readings for the same stimulus, depending on whether the stimulus is approached from above or below

icon graphic symbol that represents words and shortens the learning time for a computer application

importing transferring data from one program to another without loss of the original data or rekeying of data

integrated circuit circuit that combines many electronic components, such as transistors, capacitors, or resistors

integrating A/D A/D whose digital values represent the area under the analog signal

interdomain conversion   communication between data domains

interface   physical or software link between a personal computer and another device

interpretive language   language that enables the source code to be read and converted immediately so the instructions can be executed; interpreted BASIC is an example

justify   to make all the lines in a document the same width; the right and left margins will align

linearity   measure of how well a transducer meets an ideal calibration line or predictable function

local area network (LAN)   method of allowing a group of computers that are near each other to share information and computing resources

logbook   program that can be used to track the results of tests performed on any final or intermediate manufactured product

machine code   instructions that the central processing unit can execute; also called object code

machine language   language executed by the central processing unit

macro   tool that allows several commands or keystrokes to be executed by using a single keystroke

math coprocessor   optional chip(s) that add additional arithmetic instructions to the main microprocessor

memory   hardware in which program instructions and data are stored while the central processing unit performs calculations

modem   device that links a computer to a telephone line and allows communication with a remote computer

monitor   main device used to display entered data and results

mouse   pointing device that allows a cursor to be manipulated on a screen

network analysis   techniques that manipulate diagrams showing the relationship between tasks and activities that must be done to complete a project

object code   instructions that the central processing unit can execute; also called machine code

open architecture   capability that allows users to create add-on hardware for specific applications because the computer company publishes specifications for the add-on hardware slots

operating system   system that controls the basic operation of the computer software

paging technique that allows a central processing unit to address additional memory

painting program program that allows free manipulation of each screen pixel for image generation

parallel communication capability that allows a set of bits to arrive at a peripheral personal computer component at the same time

parallel port device that allows a personal computer to communicate with other devices by sending many bits of information simultaneously

parity method for error checking that counts the number of *1* bits in a communication

PC-DOS personal computer disk operating system; it controls almost every aspect of the basic operation of a personal computer

personal computer computer designed to be used predominantly by one person

pixel picture element or dot on a computer screen

port device that allows a computer to communicate with devices outside its own enclosure

precision reproducibility of a transducer measurement

print queue memory area in which a job is stored when the printer is printing another job

procedural language language that has commands for retrieving, editing, or storing data in data bases; any language used to extend the utility of a program

protocol formalized set of conventions used for establishing and maintaining contact between two communicating devices

pulse electrical signal that rises and falls rapidly to form a square wave form

random-access memory (RAM) memory used to store the executing program instructions and data temporarily; also known as read–write memory

read-only memory (ROM) memory used to store unchanging program instructions like those needed when the computer is first turned on

record data management term describing a data base entry that forms a relation by entering values for each field or category; also called a tuple; storage device in which fields are kept; records are located in a file

record locking method of allowing only one user in a network to change an existing record

register memory location within a central processing unit

relation table in a data model for a data base management system; also called a file

report generator the part of a data management system program that allows reports to be created, usually on the display screen, which allows the user to place information in the desired location

resolution the number of discrete values that can be assigned to the analog level

response factor factor computed by dividing a compound's detected area by its concentration to the standard in a chromatographic run

sampling rate number of A/D conversions per second that are performed

screen dump one-to-one reproduction of the screen

scrolling viewing a document or file one screen at a time

serial communication communication that allows only one bit to arrive at a time to a peripheral computer component

serial port device that sends and receives data one bit at a time; also called synchronous communications port

simplex transmission of data in one direction

software the variable part of the computer that includes the programs

source code instructions written by programmers in high-level languages

spreadsheet program that consists of columns or rows; when data is input into these columns and rows, many calculations can be performed

stop bit special character that is sent from one computer to another to indicate where a character's data ends

strobing method of sychronizing the transmission of characters

systems programs software tool that makes the computer easier to use regardless of the application program

terminal emulation program program that allows the user to set the necessary communication parameters and start communication with an instrument

text merging capability that allows the user to substitute a block of text at a specific position in each printed document

throughput rate of data collection, including all steps required to monitor and store a signal

transducer device that converts a physical stimulus into an electrical signal

transfer error error initiated by a computer interface that can usually be detected and resolved

tuple row of information in a relation in a data base management system; also called a record

wire electrical path along which data is transmitted

worksheet application of a spreadsheet program that includes calculations and report formats; by changing raw data, a new report can be generated; also called a template

Index

A

Above Board, hardware use, 27
Absolute reference
  identifying a cell identifier, 125
  specific cells in a spreadsheet, 125
  use, 91
Active talker, 208
Activities, description, 188
Ad hoc query, data-retrieval method, 157
Address, description, 24
Addressing memory, 24–27
Aldus Page Maker
  complete printed page, 83f
  description, 82
American Chemical Society, computer courses, 10
American Standard Code for Information Interchange (ASCII), code description, 26
Analog voltage-to-digital (A/D) converters
  acquisition rate, 244
  chromatographic data, 244–246
  description of use, 236–244
  digital result, 242f–243f
  dynamic range, 238
  See also Integrating A/D
Answer/DB, use, 214
APL, stand-alone statistical program, 193
Apple Macintosh
  application environments, 55
  desk-top publishing, 82

Application environments, 54–58
Application express, R:Base 5000 implementation aid, 182
Application programs
  compatibility with GEM application environment, 57
  use on multiuser, multitasking operating systems, 52
Application software, 13–16, 71
Area percent report
  computation of percent total area, 90
  creating column headings, 90
  generated in Lotus, 89f
  report header information created, 89
  use, 88
Arrow keys, use, 30
ASCII code
  character number, 203
  transmission of a character, 201–202
Assembler, use, 63
Assembly, description and ease of use, 67
Asynchronous communication ports—See Serial ports
Asynchronous communications
  comparison with parallel communications, 201
  modem, 199
ASYST
  data smoothing, 249f
  use, 247–248
ASYSTANT, use, 247–248

Protocol, definition, 198
Pulses, V/F converter use, 246

Q

Quality data charts
 data addition, 128f
 description, 124–129
 example, 100f
 graphics, 126–127f
 labels on graphed lines, 131f
 preprogrammed function, 99
 scale and range alteration, 128
 standard deviation computed, 99
 user-defined scale, 129f

R

R:Base 5000
 browse, 184
 DBMSs, 19
 delete, 184
 edit, 183
 implementation aid, 182–185
 print, 184
 select, 184
Random access memory (RAM)
 disk use, 61
 use, 23
Read-only memory (ROM), use, 24
Read–write memory—See Random
 access memory (RAM)
Reading, source of brainware, 9–10
Record locking
 definition, 219
 programs, 155
Records
 definition, 157–158
 relationship to fields and files, 159f
Reflex
 creating a data base, 162
 Crosstab view of data, 167f
 description of uses, 162–168
 editing data in the data base,
  162–165
 entering data into a data base, 162
 Form view of data, 164f, 167f
 Graph view of data, 166f
 List view of data, 164f, 166f
 main menu, 163f
 report writer, 165–168
 searching and sorting data, 165
 views of data, 165

Registers, use in CPUs, 22–23
Relation, examples, 180
Relational data base management
 systems, advantages, 178
Report generators, data-retrieval
 method, 157
Report writer, Reflex, 165–168
Reserve words
 description, 46
 list, 47t
Resolution
 A/D converters, 238
 chromatographic data, 245
 high, 152
 middle-of-the-road, 152
Response factor
 concentration computed, 92f
 definition, 91
Ring configuration, use, 220
RS–232C
 data communication, 202
 serial communications interface,
  200f
 transmission of characters, 201f
 use, 197
Rule number, use, 200

S

Sampling rate
 A/D conversion, 237
 effects on sine curve, 239f–241f
Scatter plots, STATA, 193
Scientists, graphics use, 109–110
Screen dump, definition, 145
Scrolling, use, 75
Search and replace, word
 processing, 78–80
Secondary key, defining, 174
Select, R:Base 5000, 184
Serial, definition, 198
Serial communication
 description, 27
 hardware links, 200, 205–206
 use, 199
Serial ports, use, 28
Sidekick
 description and use, 59
 menu of functions, 60f
 time-saving ability, 60–61
Simplex, data transmission, 202–203
Single-level standard quantitation,
 spreadsheet programs, 91–94

Copyediting and indexing by Keith B. Belton
Production by Paula M. Bérard
Book and cover design by Pamela Lewis
Managing Editor: Janet S. Dodd

The text was typeset in Chelmsford and
the titles were set in Eurostile by Hot Type Ltd., Washington, DC.
This book was printed and bound by
Maple Press Company, York, PA.

Recent ACS Books

Writing the Laboratory Notebook
By Howard M. Kanare
145 pages; clothbound ISBN 0–8412–0906–5

Polymeric Materials for Corrosion Control
Edited by Ray A. Dickie and F. Louis Floyd
ACS Symposium Series 322; 384 pp; ISBN 0–8412–0998–7

Porphyrins: Excited States and Dynamics
Edited by Martin Gouterman, Peter M. Rentzepis, and Karl D. Straub
ACS Symposium Series 321; 384 pp; ISBN 0–8412–0997–9

Agricultural Uses of Antibiotics
Edited by William A. Moats
ACS Symposium Series 320; 189 pp; ISBN 0–8412–0996–0

Fossil Fuels Utilization
Edited by Richard Markuszewski and Bernard D. Blaustein
ACS Symposium Series 319; 381 pp; ISBN 0–8412–0990–1

Materials Degradation Caused by Acid Rain
Edited by Robert Baboian
ACS Symposium Series 318; 449 pp; ISBN 0–8412–0988–X

Biogeneration of Aromas
Edited by Thomas H. Parliment and Rodney Croteau
ACS Symposium Series 317; 397 pp; ISBN 0–8412–0987–1

Formaldehyde Release from Wood Products
Edited by B. Meyer, B. A. Kottes Andrews, and R. M. Reinhardt
ACS Symposium Series 316; 240 pp; ISBN 0–8412–0982–0

Evaluation of Pesticides in Ground Water
Edited by Willa Y. Garner, Richard C. Honeycutt, and Herbert N. Nigg
ACS Symposium Series 315; 573 pp; ISBN 0–8412–0979–0

Water-Soluble Polymers: Beauty with Performance
Edited by J. E. Glass
Advances in Chemistry Series 213; 449 pp; ISBN 0–8412–0931–6

Historic Textile and Paper Materials: Conservation and Characterization
Edited by Howard L. Needles and S. Haig Zeronian
Advances in Chemistry Series 212; 464 pp; ISBN 0–8412–0900–6

For further information and a free catalog of ACS books, contact:
American Chemical Society, Sales Office
1155 16th Street NW, Washington, DC 20036
Telephone 800–424–6747